D0344613

The **True** Sound of Music

The **True** Sound of Music

A PRACTICAL GUIDE TO SOUND EQUIPMENT FOR THE HOME

by Hans Fantel

New York

E. P. Dutton & Company, Inc.

1973

First Edition

Published simultaneously in Canada by Clarke, Irwin & Company Limited, Toronto and Vancouver

SBN: 0-525-22380-0
Library of Congress Catalog Card Number: 72-82716

For Shea

who listens

Acknowledgments

Parts of this book are based on articles that have appeared in *Stereo Review,* where I served as associate editor for several years. Mr. William Anderson, the editor of *Stereo Review,* has kindly given permission for the use of this material.

I am grateful to Ivan Berger for assistance in adapting, expanding, and updating these articles for inclusion in this book. His help and encouragement have been truly generous.

H. F.

Contents

List of Illustrations

Preface

Listening to music has been one of the most profound and enjoyable of human experiences since the beginning of civilization. Yet in our own lifetime this experience has been totally transformed by the development of modern sound equipment. Most music heard in the world is no longer heard in the presence of the performer. Music is no longer instantly perishable. It no longer vanishes with its own sound, nor is it bound to any location.

Through radio and recording, music has become independent of time and place. The music-making of many years, from all over the world, can now be heard in your home, at any time, simply at the flick of a switch.

As historian Jacques Barzun points out, the phonograph has done for music what the printing press did for literature. It has made music a vital part of everyday life for countless people who otherwise would have had little or no access to it.

But the full pleasure and meaning of music can be transmitted only if the equipment in your home can do justice to the true nature of musical sound. You might think of your sound equipment as a means of transport that takes music through time and space and delivers it to your house. With most ordinary radios and phonographs, the music gets damaged in transit. The color and range of musical sound are falsified and diminished—the sensory and emotional impact is lost along the way. It takes carefully chosen equipment to retain the breath of life in reproduced music. In short, the electronic conveyance of music must be worthy of its freight.

The purpose of this book is to make your listening more rewarding by acquainting you with the crucial quality factors in sound reproduction. If you are interested in sound reproduction strictly for the love of music and lack a technical turn of mind, you may feel somewhat bewildered by the complexity of modern sound equipment. This book will explain the basic principles of audio and clarify the technical terms that describe the design and specify the performance of sound equipment.

I will steer clear of nuts and bolts, circuits and formulas. The object, after all, is not to make an engineer of you but to give you a general perspective that will enable you to form your own reasoned opinions of the new developments in sound reproduction, allow you to choose your equipment intelligently in accordance with your needs, and help you get the best possible performance from the equipment you own.

H. F.

A Word About Words

The first requirement for fruitful discourse about anything is that the meaning of the basic terms must be clearly understood. Audio, like any other special field, has its own terminology, since specific words are needed to designate specific concepts. Too often, however, technical language defeats its own purpose of precision and clarity by degenerating into obfuscating jargon. I have made efforts to avoid this particular pitfall.

This book sticks to common language as far as possible so that it can be read by anyone without any prior knowledge of the technical aspects of sound reproduction.

However, as technical concepts are introduced, technical terms for them necessarily enter into the discussion. Each is explained the first time it appears. Occasionally you may have forgotten the definition of an unfamiliar term when you encounter it later on. In that case, turn to the Glossary at the end of the book, where essential terms are alphabetically listed and briefly explained.

The **True** Sound of Music

1
Mono, Stereo, and Quad—Three Perspectives on Sound

There are three basic kinds of sound equipment available today: (1) monophonic, (2) stereo, and (3) quadraphonic four-channel systems, sometimes called quad for short. This chapter deals with the difference and relative merits of these three types of systems.

Essentially, the difference lies in the number of sound channels used in reproducing the music. Each channel represents a separate signal source, a separate amplifier, and a separate speaker. Monophonic equipment (mono for short) uses one channel only. It is found mainly in portable radios where everything has to be in one piece. In home-based, quality sound equipment, mono is no longer a major factor. The current standard for quality sound reproduction is stereo, which uses two separate channels.

A flurry of excitement has been stirred up by the development of four-channel systems. At its best, quadraphonic sound yields the most realistic music reproduction available at the present state of technology. However, many low-cost quadraphonic systems fall far short of the optimum and rely mainly on the novelty of four-channel sound for their appeal. Moreover, the technical complexity of four-channel installations may limit their attraction to those hobbyists who are seriously interested in the ultimate technical possibilities of sound reproduction. For listeners mainly interested in music, and leery of technical elaborations, standard stereo may remain the preferred format.

The relative merits of these different approaches to sound reproduction are best understood in the historical perspective of their development. A decade or so ago, many audio fans were skeptical about an innovation called stereo. Their main objection was that stereo required two speakers, two amplifying channels, and replacement or conversion of the tuner, record player, and tape recorder. To many people who had been listening happily to their mono systems all along, stereo seemed an opportunistic maneuver by the audio industry to double its business. Their suspicion was reinforced by the fact that the early stereo discs sounded none too impressive, being replete with sonic and spatial distortions. But when these technical problems were finally solved, it was evident that stereo represented a legitimate advance well worth its price.

Today audio is at a similar juncture. The first four-channel components for the audiophile market are being introduced, and records and prerecorded tapes for four-channel reproduction are available. Again the thought intrudes that the industry may be trying to treat itself to another 100 percent hike in hardware requirements; the music-oriented consumer may ask if we are dealing merely with a laboratory curiosity of interest only to the most intrepid hobbyist, or with a portent of a basic change in our music media.

It is not surprising that the questions are the same, for both stereo and four-channel are attempts to reproduce the spatial aspects of sound, its directionality and ambience.

THE AMBIENCE OF SOUND

Ambience refers not to the sound of the music created by the musicians, but to the sound of the hall in which the music is played. It is the total pattern of the sound reflections and reverberations in the hall. This ambience is so distinctive that it is possible to tell from the reverberant evidence whether a particular recording was made in the Vienna Musikverein, New York's Carnegie Hall, or Amsterdam's Concertgebouw.

Some of the distinctions are audible even through the single channel of a monophonic record. But where mono barely lets

Pianist Artur Rubinstein recording with the Philadelphia Orchestra. The microphones are, in effect, the ears of an audience extended in time and space. (*Photo: RCA*)

you detect the ambience, stereo gives it more precise definition. With stereo, you can sense the directions from which these reverberations impinge. And from these clues, your sense of hearing reconstructs a realistic image of the space in which the music was recorded. Since that recorded space is almost always larger than your listening room, stereo makes your room sound larger. When you close your eyes and listen to a stereo recording on a good sound system, your walls seem pushed back, so that the concert hall space "grows" within your living room.

It is easy to demonstrate the difference for yourself. Simply switch a stereo system from stereo to mono while listening to a stereo record or FM broadcast. Suddenly, the music—and the room—will seem shrunken, without depth or spaciousness. Stereo, in other words, adds a sense of depth, a third dimension not present in single-channel listening.

We hear depth in pretty much the same way we see it. The reason our eyes and ears come in pairs is not just to furnish a spare in case one gives out but rather to make sight and hearing three-dimensional. When you look at an object with both eyes you see it in a perspective that is lacking from any one-sided view. When you look through a stereoscopic photo viewer, figures and objects seem to stand out bodily against the background. Close one eye, though, and everything goes flat, just as sound does when you switch from stereo to mono.

To prove the importance of stereo vision, take a pencil and try to touch its point to any letter in this sentence, first with both eyes open, then with one eye closed. You can do it, but you will find it much harder to hit the right spot without the stereo space sense of normal, two-eyed vision. This space sense derives from the fact that each eye literally has its own point of view. From the difference in the lines of sight between the two eyes, the brain computes what we experience as three-dimensional perspective. A similar process applies to two-eared hearing. Similar—but not identical. For the sense of directionality in hearing is different in some ways from directionality in seeing. Close one eye, and you have no more trouble distinguishing the direction of an object than you did with both eyes open. But cover an ear, and your sense of sound direction almost disappears.

Each of your ears points in a different direction, so your two ears never hear exactly the same thing. Suppose a car honks at you as you cross the street. Your head jerks toward the sound. How did you know which way to look? The ear aimed toward the car heard the horn more loudly than the other. It also heard the horn sooner, for the sound takes time to travel around your head. From these tiny differences in loudness and arrival time, the brain computes the approximate direction and distance of the sound source. That is what we mean by stereo hearing.

Checking his recording in the control room, Rubinstein advises engineers on tonal balance and nuance. (*Photo: RCA*)

At a concert you can spot the general position of the players blindfolded. Even with your eyes closed, you can follow what is going on in the orchestra—fiddles left, cellos right, winds in the middle, and percussion at the rear. Musically, that is important. For such stereo hearing makes each strand of melody stand out against the background, clears up the different tone odors, and lets the music come alive as it never could in one-eared hearing.

"Well," you say, "I always listen with both ears. So what's the point?" The point is that listening to monophonic (non-stereo) radios or phonographs is like listening with one ear, because in mono you have only a single channel of sound transmission, giving you only a single signal. It is not like natural hearing, where each ear picks up a slightly different sound to give you a sense of sonic perspective. To get the third dimension into reproduced sound, you need two separate signals, therefore two separate transmission channels.

Stereo equipment provides just that. The second channel is the electronic equivalent of your "other" ear. So, in effect, stereo puts *both* your ears in the studio or concert hall. In short, stereo incorporates into sound equipment the same two-channel principle that nature designed in our heads. That way, stereo records and broadcasts bring you the sound of music in its natural space dimensions.

But if two-channel stereo does all that, why try for four channels? Especially if we only have two ears? The reason is still the same: accuracy of sonic perception. We can get more from four-channel sound than from standard stereo because in a sense we do have more ears than two. While we do not have eyes in the backs of our heads, our ears can in fact perceive "backward." Much of the total sound energy you hear in a concert hall actually does come from behind you. This is the sound bouncing off the rear and side walls of the hall. Because the reflected sound path (via rear-wall bounce) is longer than the direct sound path from the front, there is a time difference between what you hear from front and back. It may just be a couple of milliseconds, but this tiny time difference gives your brain vital clues that help it sense the space in which the music is played—ambience again.

Conventional stereo catches some of this "reverb" from the rear. But it all emerges from the speakers up front. The idea of quadraphonic sound is to distribute the reverb between front and rear to create an all-around sound field, the way it happens in a "real" listening situation.

It works. I have heard symphonic music played in the four-channel mode on top-rated equipment, and the result is stunning. I simply forgot where I was. I felt that I was sitting at the acoustic focus of some ideal concert hall.

Such realism represents the approach to four-channel recording used for classical orchestral music. The approach more often used in quadraphonic recordings of popular music is to give the listener the feeling of being in the *middle* of the music itself. There you gain the impression of sharing the stage with the musicians, sitting among them, rather than sitting at some distance in the auditorium. Done well, this is a very dynamic, absorbing presentation of sound that intensifies your emotional involvement in the music.

ALTERNATE SOUND PRESENTATIONS

To clarify the difference between the sound impressions gained from mono, stereo, and quad, let us compare monophonic listening to hearing the music from a seat in a room with an open "window" into the concert hall. Stereo, in effect, knocks down the wall between your listening room and the concert hall. Then, going one step further, quadraphonic sound puts you, for the first time, within the hall itself.

Four-channel sound naturally requires four speakers and four separate reproducing chains to feed each speaker a separate signal. Inevitably, this doubling of channels increases the cost of four-channel sound systems. The price tag of a quadraphonic system won't be twice as much for a roughly equivalent system, but it will be at least 50 percent and perhaps as much as 70 percent more. So the question arises: Is it worth the extra outlay?

This is a question everyone must answer for himself, depending on his personal impression of quadraphonic sound and the subjective satisfaction he derives from it. My own feeling is that four-channel provides a distinct margin of sonic experience but not a decisive one. In terms of my personal response to music, I find that quad—in comparison to stereo—does not add as much to my enjoyment of music as stereo does in comparison to mono. After all, it does not provide inherently truer sound than stereo; what it offers is an expanded sense of aural ambience.

To discover for yourself the difference between quadraphonic and stereo sound, you can make the following simple experiment. Ask your audio dealer to demonstrate a four-channel system to you. Then, as you listen, switch to two-channel stereo. Every four-channel system has provisions for such switching, so you can alternate between quad and stereo as the music plays. (Of course, while listening in the two-channel mode, the rear speakers should be muted.) Once you hear the difference in such direct, side-by-side comparison, you can decide for yourself whether the sonic augmentation of quad is worth the added cost. If you like the music almost as well in two-channel stereo, there is little reason for you to spend the extra money for quad.

Looking at the economics of the situation another way, you can buy a superior stereo system (superior in terms of amplifier power and speaker capabilities) for the same money you would spend on a quadraphonic system with less capable speakers and amplifying channels. With the audio industry pushing four channels in every segment of the consumer market, product designers are apt to skimp on such basic requirements as speaker range (particularly in the low bass) and power per channel, so as to make these systems competitive in price with standard stereo. The inevitable compromise is in fidelity. After all, sufficient power and range are still the prime requisites for good sound. Neither of these should be sacrificed merely for the sake of having four sound sources instead of two. If you can afford four top-ranking sound channels and if you find significant personal pleasure in its sonic ambience, a four-channel system will be a rewarding venture for you. But remember that two good sound sources will give you more pleasure than four mediocre ones.

"But if I buy a stereo system now, won't it become obsolete as four-channel gains increased popular acceptance?" This question is heard in audio shops more and more often as prospective buyers try to make up their minds between stereo and quad. Essentially, it is a moot point. A system does not become inherently obsolete. If its sound gives you pleasure now, it will continue to give you pleasure. No matter what technical developments become available in the future they will not make your system worse. Its musical capabilities will remain the same. What's more, there is the possibility of updating a stereo system to provide four-channel capabilities later on. Chapter 8, which describes the actual hardware, will discuss this point in detail. For the present it is enough to know that whatever four-channel formats may evolve in the future, they will be compatible with standard stereo. While engineers differ about the best way to achieve four-channel sound on records and via FM broadcasts, they all agree on one point: they do *not* want to make standard stereo obsolete. The possibility of your wanting quad at some time in the future therefore need not deter you from buying stereo now. Think of quad as an added option for stereo, not as a replacement.

2
Consoles, Compacts, and Components— Three Basic Options

When you make your first trip to an audio showroom, it helps to know in advance just what you want. We have already discussed the choice between mono, stereo, and quadraphonic sound. Now you are faced with a further choice between three basic kinds of sound equipment: consoles, compacts, or components. This chapter spells out your basic options, detailing the advantages and drawbacks of each.

A console system is entirely contained in a single piece of furniture. A so-called compact music system has independent speakers connected by long wires to a central unit which contains everything else. A component system consists of separate, functional building blocks—speakers, amplifier, turntable, and tuner—each of which may be from a different manufacturer.

My own preference is for components, for reasons explained later. But there are some things to be said in favor of the other two types of systems.

THE CONSOLE: THE MODERN MUSIC BOX

The console is the most deliberately decorative of hi-fi systems and the one that most obviously affects your room's appearance. It is also the simplest to buy and install: just plug it in, and your room is filled with music, as well as with an outsize piece of furniture. But its advantages, alas, end there. For the emphasis on furniture affects both cost and quality of consoles. The fancy cabinetry alone may well account for more than half the console's price. There is nothing wrong with spending money on furniture, but you should know where your money is going when you buy a console. It goes mostly for the box, not for its contents.

Since many console buyers are as much interested in furniture as in sound, console makers tend to put their emphasis accordingly. Often more concern is given to the cabinet than to what it contains.

A typical *de luxe* stereo console. The emphasis is on the cabinetry.

Not only does the cost of the cabinet take up most of the price, but reputable makers must spend still more of the total cost budget to overcome a technical problem inherent in console design: acoustic feedback. Acoustic feedback occurs when the bass notes from the speakers shake the stylus in the record groove. The amplifier then reamplifies these bass notes, feeding them to the speakers again—and louder. This shakes the stylus still more vigorously, and the cycle repeats itself over and over. This adds an annoying boomy quality, excessive "bass." In occasional extremes at loud passages, this feedback can become an ear-splitting, exasperating howl. (The British, in fact, call it "howl-round.")

The obvious solution would be to move the turntable away from the speakers. But since in consoles the speakers and the turntable are confined together by the cabinet, this is impossible. And since the cabinet carries vibration directly from the speakers to the turntable, a cure is desperately needed. There are two possible solutions to this problem—one elegant and expensive, the other cheap and dirty. The elegant way is to isolate the turntable, the speakers, or both from the cabinet itself by padding or a spring suspension. The other approach, which is a great deal cheaper, avoids the problem altogether; if low bass can shake the turntable, eliminate the lower bass notes. As long as the bass gets "thrown away" the manufacturer also can use cheaper speakers and skimpier cabinetry, thus saving money all around. (A quick check: if the speaker enclosures have no solid backs or bottoms, or widely slotted ones, it is a sure sign that the manufacturer has been cutting corners.)

Speaker placement is another problem with consoles. For proper stereo, speakers should be at least eight feet apart in average-size rooms and as much as twelve feet apart in large rooms. With both speakers built into a single console, the cabinet would have to be anywhere from nine to thirteen feet wide. Yet most consoles are only about six feet in width. The compromise is obvious.

Even if the console's width were adequate, speaker placement would remain a problem. Speakers do not always sound best wherever the console fits into the room. The console's dominating bulk limits choice of placement. By contrast, separate speakers can easily be placed where they sound best.

In addition to their inherent drawbacks, consoles are inconvenient in use. Controls are inaccessible when you are sitting in the best listening area, which is across the room from the speakers. Besides, the pace of technical advance and the accelerating rate of furniture style changes may soon make either your console or its contents outdated. So if you intend to update your system or redecorate your room periodically, you would do better with a compact or component system, in which separate parts can be exchanged.

Nevertheless, if your taste or your decor demands a console, it is still possible to end up with good sound. Several component manufacturers produce consoles that are actually component systems grouped together in a box. But with a typical console you take pot luck with whatever the manufacturer has hidden away in the furniture.

Unfortunately, reliable performance specifications are not always available for consoles. Most manufacturers provide only a promotional blurb without numerical specifics. Even when console makers go through the motion of listing performance data, their statements are often calculated to impress rather than inform. A case in point is their penchant for rating consoles by "EIA music power," or sometimes even "peak music power," instead of the more accurate IHF music power or continuous (RMS) power ratings. Differences between these rating systems will be discussed in Chapter 5. For the present, I should only point out that these confusing measurements standards allow a 20-watt amplifier (respectable, if unspectacular) to be advertised enthusiastically as offering "100 watts of room-filling, ear-shattering power."

The best way to buy a console is to buy separate components

and have them installed in a console cabinet that pleases you. Cabinets for just that purpose are available at better audio salons. You may lose some of the console's inherent simplicity, but you will more than make it up in sonic excellence.

THE COMPACT: GOOD SOUND ON A BUDGET

The compact music system shares the console's basic advantage: the simplicity of the package concept. No agonies of choice over which turntable, which cartridge, which amplifier, which tuner, or which speakers to combine. A compact is virtually as easy as a console to install. If a console's installation is as simple as plugging in a lamp, setting up a compact is like plugging in three lamps: the power cord that plugs into the wall and the speakers that plug into the central unit. That's all you have to do. It only takes a minute to complete the hook-up—and you're ready for the music. This ease of installation is a boon to people who are naturally wire-shy.

Unlike the console, the compact represents the optimum cost-performance ratio. It gives the best possible sound for the money. Surprisingly, the compact's economy and the console's extravagance have the same basic cause: the cost of cabinetry. The console cabinet must stretch wastefully across the entire space between the speakers. Wrapping that space in wood costs money. Justifying the space by filling it with record racks or a bar costs still more. The compact, by contrast, houses the central equipment in a simple, space-conserving cabinet that fits onto a table top or bookshelf—a drastic saving in space and cost.

Compacts suffer from none of the technical problems and limitations of the console. The speakers, in their independent cabinets, can be placed wherever they sound and look best; the central unit can be placed wherever its controls are convenient; and feedback from the speakers will not shake the stylus in the groove, because turntable and speakers are not in the same cabinet.

Two typical compact systems. Record player and electronic equipment are combined in the center unit. Loudspeakers are separate. (*Photo: Electro-Voice and Scott Corp.*)

Compacts offer nearly the same versatility as separate components. Virtually all compacts have input and output connections for tape recorders, and many provide output jacks for a second pair of speakers to play in another room. They present just one problem: sometimes it is hard to distinguish them from ordinary table-model phonographs. The latter, for the most part, are nothing but ear-grating compendia of technical shortcomings. They chop off all the lower notes, leaving the music shrill and thin (surprisingly so, considering how brutally they also chop off the extreme high frequencies). The aural solidity and tone so essential to the reproduction of orchestral music are utterly beyond the scope of their capabilities.

How then do you tell an honest, capable compact system from the ordinary kind of table-top stereo phonograph? To start with, check the specification sheet, the basic performance guide that will be discussed on later pages. And, as a rough index of capability, you can check whether the speaker enclosures have open or slotted backs. If so, dismiss the model as inferior. Only fully enclosed loudspeakers with solid-wood back and bottom have acceptable sound. Finally, you can get a pretty good idea of a compact's basic merit from its brand name. If it is made by a manufacturer normally identified with ordinary home entertainment products (television, consoles, portable or kitchen radios), chances are that his compacts are minimally designed. On the other hand, if the manufacturer is also known to make high-quality audio components, his compact systems will follow more exacting engineering standards. Among the top-rated compact manufacturers are such renowned names in the audio field as KLH, Scott, Fisher, and Harman-Kardon.

COMPONENTS: THE WAY TO THE ULTIMATE

For all the virtues of compact systems, I am still partial to components. No other type of system offers the same flexibility and adaptability, nor can you approach ultimate sound quality by any other route. For what they offer, components are not really expensive. They cost but little more than compacts of comparable quality and considerably less than consoles. They are nearly as economical of space as compacts and more flexible in the way they can be set up on your shelves.

Paradoxically, components offer good value for the same reasons they are regarded as luxury items. Sound quality is the component maker's *raison d'être,* so he cannot afford to cut corners on quality if he wants to retain his special market. Recognizing this, most component manufacturers structure their cost budget differently from makers of other types of home electronic equipment. They keep corporate accountants out of the engineering department and snip parasitic links from the distribution chain. To offer the quality demanded by their "golden-eared" clientele while staying competitive in price, some component makers and their specialized retailers operate on narrower profit margins than their counterparts in the furniture and appliance businesses. Components thus offer the customer a price-performance ratio hardly approached by any other line of technical consumer goods.

A component system in a space-saving wall installation. (*Photo: 3-M Corp.*)

More significantly, a quasi-artistic tradition still prevails in many parts of the component industry that aims, above all, to make the product a fit conveyance for the true splendors of music. Thus, the most significant refinements of circuits and performance appear first in top-rank components. But as soon as possible these advances filter down to components in the medium-price bracket. In consequence, components are usually in the technical forefront, far ahead of the other kinds of equipment in sophistication and capability.

With some knowledge of the basic audio principles, as set forth in this book, and with judicious use of your ears, you will almost certainly find truly satisfactory models among the wide choice of components on the market. Even if you do become dissatisfied with some part of your system (more likely due to growing sophistication of your ears than to inherent defects in your equipment), only components let you replace one part of your system without losing any of your investment in the others. And you don't have to worry that some of your component choices will not work properly with the others. Nearly all makes of components are compatible with each other. Incompatibility is a myth, fostered largely by advertising talk of "matched" components, that is, components from a single manufacturer who naturally wants you to buy all items from his own line of products. The few areas in which a mismatch is possible can be checked out in advance at any dealer's showroom.

The perfect match that packaged systems offer you is one of quality level, not of design compatibility. A medium-quality compact, for instance, will include a fairly good turntable, a fairly good tuner, fairly good amplifier, and fairly good speakers; a more expensive system will be uniformly better all across the board. After all, there is no point putting a Cadillac engine in a Volkswagen chassis. Most component buyers match their systems up precisely the same way.

But one of the great advantages of component systems is that you can deliberately mismatch your components to create a system that better matches your particular individual needs. If you want to fill a big room with the loudest possible rock sound or symphonic music, you may choose a more powerful amplifier than would the person whose room is small and whose tastes run to string quartets. Or if you listen mainly to nearby FM stations, you can economize on your tuner. But if you live far from your favorite FM stations, you can buy the best available tuner to pull in those distant signals, yet keep the rest of your equipment at a modest cost level. In either case, separate components let you pair any tuner with any amplifier, to come up with a combination to fit your special requirements. Moreover, should your requirements change—should the chamber-music aficionado find that his children have a taste for rock, or the city-dweller move to an FM fringe area—or even if you wish merely to upgrade your system to match improvements in technology (or in your financial status), you can adapt your system by trading in just one component at a time. This gives you a greater choice in updating and adapting your system.

Component flexibility also lets you add extras to your system gradually, as your interest or your cash reserves dictate. What's more, you can add these extras in virtually unlimited variety. Most compacts offer only one extra set of auxiliary connections (consoles rarely offer even that). But the rear panels of even the most modest amplifiers are crowded with terminals for connecting tape, cassette, and cartridge units.

As mentioned before, both components and compacts let you place your speakers where they sound best in your particular room's acoustics and far enough apart for optimum stereo. At the same time the controls can remain within reach of your favorite chair, so you don't have to jump up and run across the room every time you want to change the volume or the station you're listening to. Should trouble arise, it can be located easily in one of the separate components. These individual units are then easier to carry to a repair shop, and because components may be repaired separately, you retain the use of the rest of your system while the ailing component is out being fixed. For example, if your FM tuner needs fixing, you can still play records, thus avoiding total music deprivation. And because their circuits are more accessible, components usually cost less to fix than do equivalent compacts or consoles.

Despite all these advantages, some buyers shy away from

components, either because they fear the complexities of hooking up their systems or because they envision a component system as an unsightly maze of connections, gadgets, and wires. The hook-up problem need not deter you. Most dealers will install a system for you. And every neighborhood, office, or country club boasts at least one amateur audio expert who can help you. In fact, installing the system yourself is really not difficult, unless you've brainwashed yourself into believing that such tasks are beyond you. Strict attention to the printed instructions provided with each component will nearly always guide you adequately. And you'll find general guidelines and helpful tips in Chapter 12 of this book.

The charge that component systems are unsightly was true only in the early days of audio. Today components are styled to please the eye as well as the ear. You can leave tuners, amplifiers, and turntables sitting "naked" on a shelf. They'll look good, blend with almost any decor, and—in contrast to bulky consoles—take up no floor space whatever. Today's components obviate the need for cabinetry; they let you put your money where the sound is.

Receivers

The latest trend in component design is to combine tuner and amplifier in a single unit called a receiver. This is a logical outgrowth of the latest advances in solid-state electronics. Modern techniques of solid-state miniaturization now make it possible for engineers to cram all electronic parts of a sound system into a single unit the size of a shoebox. Many of these receivers now do a better job of sound reproduction than separate tuners and amplifiers did only a few years ago. Thanks to their compactness, receivers have become the most popular type of component, being generally preferred over separate amplifiers and tuners. Yet the latter are still widely used in high-powered installations, where the need for hefty wattage makes compactness more difficult to attain.

Receivers offer other advantages aside from compactness. As a rule, they cost less than separate amplifiers and tuners of similar quality because the combination of all parts on a single chassis permits some cost-saving in manufacture. Moreover, the combined tuner and amplifier sections can share the same AC power supply. Aside from this, receivers are simpler to hook up than other components. All you have to do is plug in the turntable and connect the speakers, and the whole system is ready to play. Finally, having just one attractively styled unit on your shelves instead of separate tuner and amplifier makes the equipment less obtrusive in your living room. In sum, receivers satisfy even the exacting requirements of component buyers and offer a convenient and practical way to acquire a component system.

SHOPPING STRATEGY

No matter what type of sound equipment you finally decide on, you can benefit from certain shopping strategies that will protect you against shoddy products and unscrupulous merchants, both of which form a rather unsavory substrate in the audio trade. This is the more regrettable since the majority of dealers are honest and as genuinely concerned about the satisfaction of their customers as are the manufacturers of quality equipment.

Like a bartender or a lawyer, a high-fidelity dealer should be something of a personal confidant. Tell him about your musical tastes, let him know whether you mostly listen alone or in company, describe to him the layout of your home, and be frank about your finances. Such information helps the dealer steer you toward just the kind of components to fit both your needs and your budget.

A good audio dealer can sharpen your appreciation of the important differences among various components and help you focus on factors that really matter in terms of your particular requirements. One quick way of sizing up a dealer's competence is to ask for an explanation of some of the technical points treated in the preceding chapters. If the dealer's reply is patient and plausible, to the point, and free of irrelevant sales pressure —if he seems genuinely concerned about *your* understanding— you are very likely dealing with a man who is in the audio business because he himself cares greatly about good sound and wants others to care about it. He is one of the many high-fidelity dealers who turned their hobby into a profession. Him you can trust.

This type of dealer is usually found in the specialized audio salons that now exist in nearly all larger cities. Often you can locate them through the Yellow Pages of the telephone book under such headings as "Audio" or "High Fidelity." If you are a relative newcomer to high fidelity, it is worth your while to seek out such specialized audio shops, for there you will get unhurried attention, a chance to establish personal rapport with the dealer, an opportunity to compare many different types of components in a pleasant, homelike atmosphere. Many of these dealers discuss decor with you. Some will find it easier to help you if you bring along a sketch of your listening room, showing size and proportions, the location of windows and doors, bookshelves, tables, couches, and, above all, the location of your favorite chairs. This will help the dealer plan your sound system to harmonize with the room and also suggest the best acoustic placement of the loudspeakers. Be sure to let him know if you would like to have components set into furniture you already own. Or you may discuss the possible arrangement of components as wall units, room dividers, shelf installations, perhaps even the mounting of components in a closet door, which is an inexpensive and space-saving way of housing them. However, only a few audio dealers provide such extended service.

Some of the larger discount houses and electronic-parts dealers also maintain audio departments in which components are sold, often at prices from 10 to 20 percent below those charged in the specialized high-fidelity salons. But these savings must be balanced against the fact that few such outlets offer the patient personal consultation that is the hallmark of the specialized dealer. Nor do they, as a rule, offer to help you set up your equipment in your home or provide a store warranty to assure proper operation of the whole system once it is set up. (Of course, each separate component is guaranteed by its manufacturer.) If you are a relative newcomer to high fidelity, the extra service offered by the specialized dealer may well be worth the slightly higher price.

By all means, avoid those catch-all trade emporia that sell all kinds of marked-down merchandise from washing machines to kitchen clocks, with high fidelity (sometimes of doubtful brands) thrown in on the side. There are no special bargains in high fidelity, for—unlike the prevalent practice in the home-appliance field—the nationally advertised prices of components represent realistic fair value that leaves little room for catch-as-can discounting. When you buy components, you get a bargain without bargaining.

Nor is it a good idea to shop for components in small neighborhood TV service shops that sometimes sell a sideline of audio. The selection available in such places is usually too limited to give you a representative choice. Hence you lose one of the great advantages of components: the opportunity to pick from a great variety of equipment that which best suits your budget and your needs. In that respect also, the specialized high-fidelity salon is your best bet.

Price Categories

It saves both time and confusion if you tell the dealer right at the outset how much money you can spend. In terms of cost, component systems can be divided into three broad classes: economy, solid middle, and deluxe. About $250 is the rock-bottom price of admission to the delights of component stereo. In that category a component system is apt to sound infinitely more musical than ordinary consoles selling at double the price. True, such a system may lack the ultimate in range and power, but—especially in apartment-size rooms—it will serve as a thoroughly enjoyable source of realistically reproduced music.

There is nothing middling about the "solid middle" class of sound systems bearing a price tag anywhere from $400 to $750. Components in this range satisfy the demands of the most richly orchestrated score and the most discriminating ear. Even the most critical audiophile would be hard-pressed to find anything essential lacking in the performance of such a system, especially in the upper region of this price bracket. You get full-range frequency response, extremely low distortion, virtually silent background, ample power even for large rooms and thundering musical climaxes, and convenient and highly versatile controls. For the majority of serious listeners, this quality level represents the highest returns in terms of listening pleasure for each dollar spent. Beyond this level of price and performance, possible improvement is very slight and very expensive.

But if you can look unflinchingly at a price tag with four figures, you may enter that arcane area of audio where nothing matters but the dedicated pursuit of perfection. Actually, the lower border of this musical elysium overlaps into the three-figure range at about $800. The other border is beyond the horizon. Let us assume that, being in such rarefied financial brackets, you have a spacious living room in which the splendors of such a sound system can be heard to full advantage, for a large room invariably improves the overall sound and particularly aids in effective bass projection. What you then get for your money is an added sweetness of sound, a subtler transparency of texture, an effortless authority in orchestral climaxes, and a feeling of unrestrained "openness" in sound that one critic has described as "sonic bloom." Moreover, you get the satisfaction of knowing that your equipment represents the ultimate frontier of the audio art. And if music is truly your passion you may feel that the closest possible approach to perfection is in itself a pleasure worth the cost.

However, in this rarefied realm of the ultimate, you run into diminishing returns on your investment. A $1000 system doesn't sound twice as good as a $500 system. In fact, as we have pointed out, the gain in quality, though noticeable, is apt to be marginal. It is in the "solid middle" price range that the relation of cost to performance is at its optimum. Incidentally, the price estimates were calculated here for systems minus tape recorders, that component being regarded as an optional extra for purposes of budgeting.

Specification Sheets

Regardless of price tag or dealers' recommendations, you can pretty well judge the objective value of audio equipment by referring to the specification sheets nearly always furnished with reputable components. They list the technical performance data of the equipment, though these data are not always complete or presented in unequivocal terms. But if you know how to interpret these spec sheets—both for what they say and what they don't say—they provide (next to your ears) the best possible guide for your buying decisions. The following chapters will therefore frequently refer to these specifications and discuss the concepts that will enable you to evaluate equipment in terms of these technical data.

3
Myths, Facts, and Theory— Concepts of Fidelity

Up to this point we have been talking about various kinds of sound systems as a whole. Before going on to a discussion of individual components, such as speakers, amplifiers, and tuners, it is worthwhile to consider some of the basic factors in sound reproduction. This means that we have to understand the physical foundation of music: the nature of sound.

Those readers who find this brief detour into theory too abstract can just skim this chapter or skip it altogether. But those who stick with it will learn the exact reasons why good equipment sounds like music and bad equipment sounds like something else. In this chapter we shall consider such basic theoretical concepts as frequency response and distortion. Since these terms invariably appear in the specifications describing the performance of sound equipment, understanding them will provide you with a realistic and critical standard of judgment.

Furthermore, acquaintance with these physical factors will give you a keener appreciation of that ultimate audio component: the sound of music.

Shakespeare unwittingly formulated a working definition of fidelity in sound reproduction: " 'Tis, as it were, to hold the mirror up to nature." For at its best, high-fidelity sound equipment is indeed a true mirror for music, reflecting the true sonic character of the original performance. The object of fidelity in sound reproduction, in essence, is to preserve and transmit all those elements that make music a pleasure to the ear and a nourishment to the spirit.

Ever since electronics has become music's willing and versatile servant, the striving for tonal fidelity in music reproduction has been a continuing quest. This particular journey has not been without its share of commercial detours, dictated by market considerations rather than musical values or genuine technical improvements. Commercial pressures in the highly competitive sound industry have often resulted in equipment designed more with an eye on profit than an ear for sound. This has beclouded the real issues in a field that abounds in more popular misconceptions than almost any other consumer product area. On occasion, false notions about value in sound equipment have been deliberately planted by manufacturers in their advertising campaigns, though this is not the practice of reputable firms. For the most part, the welter of popular misconceptions arose from the well-intentioned efforts of advertisers to describe the genuine merits of their product in radically simplified terms for the benefit of a technically uninformed public. But no matter how harmless or how good the intentions, whenever complex technical matters are reduced to simplified explanation in limited space, ambiguity and outright misinformation are inevitable.

WHAT IS "HIGH FIDELITY"?

At the outset, therefore, it is helpful to clear up a few of the myths that still cling to the topic of high fidelity and stereo. It has been a long way from Edison's brass horn to the touted virtues of modern multi-channel sound. For our purposes, it is best to pick up the trail in the early 1950s, when the concept of "high fidelity" first became popular. In theory, high fidelity denotes equipment that comes reasonably close to the true sound of music. But the term high fidelity, not being legally defined, has been used so loosely to promote inferior products (such as ordinary table-model radios or portable phonographs) that it soon became meaningless as a yardstick of quality. In fact, the words "high fidelity" embossed in gold script on a phonograph cabinet became the hallmark of shoddy merchandise. To distinguish quality sound components from mere pretenders, the buyer had no choice but to acquire some personal expertise.

A decade later, around 1960, the confusion was compounded by the advent of stereo. Even today, customers in audio shops still occasionally ask: "Is stereo better than hi-fi?" Such questions show that many people willing to spend a lot of money for the pleasures of good sound are still unclear about reliable ways of getting what they want. Today, the introduction of quadrasonic sound has added even further to the befuddlement of the buying public. As so often happens in situations in which advertising and promotion are involved, language obscures fact. To clarify the confusion, remember: terms like "hi-fi," "stereo," or "quadraphonic" provide no clue whatever to the quality of the equipment.

Nowadays you can walk into most appliance stores and take home something called a "stereo" for less than $100. And it would fairly well reproduce the spatial illusion we call stereo. But aside from that, cheap phonographs or radios usually sound shrill, and they grossly falsify the tonal color of voices and instruments. The music coming from the two speakers does give you some sense of depth and directionality. And the mere fact that there are two speakers and two amplifiers at work does make for some audible improvement over a comparable-quality mono set. But stereo alone cannot make up for other shortcomings in inferior equipment. With low-fi stereo there is usually no bass to give the music any fullness or warmth, or if there is bass it has a barrel-like boomy quality. The highs may be missing or distorted, the middle frequencies raucous, and the overall character of the instruments falsified and blurred. The sound can be as bad as that—and still be stereo. For stereo, after all, means only that two separate sound channels are being employed. Remember that the term indicates nothing whatsoever about the *quality* of the equipment. To do justice to the original musical sounds, stereo must first meet basic high-fidelity standards. Stereo, important as it is in making reproduced music more realistic, is merely a "plus" factor, something added to high fidelity.

This faces us with the question: What is high fidelity? Unfortunately, no one has ever precisely demarcated the point at which fidelity changes from low to high. Yet one can list three general requirements for faithful sound reproduction:

1. Wide-range, uniform frequency response. This simply means that the unit must reproduce the whole range of musical sound from the lowest to the highest notes, giving all their proper due without unnaturally suppressing or emphasizing any of them.

2. Low distortion. That is, the sound should not be gritty, fuzzy, or unclear, and it should be true to the characteristic tone color or timbre of each voice and each instrument.

3. Low noise. There should be no audible hum or background rumbling.

Only to the degree that these three basic requirements of high fidelity are met can stereo be musically—and technically—valid. The problem is to translate these abstractions into actual hardware in your living room. To define the physical factors affecting the quality of sound equipment, let us consider these three basic requirements in relation to the raw material of music, the nature of sound.

FREQUENCY AND PITCH

The concept of frequency is basic to the nature of sound. From one point of view, sound may be considered a purely subjective phenomenon, a sensation experienced principally by the ear and to a certain degree the entire body. But to understand the technical aspects of audio, we need to relate this subjective experience to the corresponding objective physical events that produce it.

Sounds are simply changes in air pressure brought about by the vibrations of an object—as for example a violin string, a vocal cord, or a drumskin. The air adjacent to these vibrating elements is driven back and forth, thus bunched up into a series of air pulsations. These are felt by the eardrums and perceived as sound by the brain.

If the rate of vibration is regular—that is, if each oscillatory cycle is accomplished in the same length of time—the sound will have a definite pitch. The higher the rate of vibration, the "higher" the pitch will seem to the human ear. For example, the A string of a properly tuned violin swings back and forth 440 times per second. The note it produces, A above middle C, is said to have a frequency of 440 Hz. (Hz stands for "Hertz," the standard unit of frequency, named after Heinrich Hertz, the German physicist who in 1887 achieved the first radio transmission. In the past several years, Hertz has replaced the formerly used abbreviation cps, or cycles per second.)

The first clues to the relationship between pitch and frequency were discovered more than two thousand years ago by the Greek philosopher Pythagoras, though he lacked the technical means for counting vibrations. He found that when the length of a vibrating string was halved, the pitch rose by an octave. Translating this into frequency terms, we say today that the interval of an octave corresponds to a frequency ratio of 1 to 2. Other specific frequency ratios govern the relationships of tones within an octave, although the notes within the modern well-tempered octave are not evenly spaced.

The lowest notes perceptible to human hearing lie in the range between 16 and 20 Hz, such as in the lowest rumblings of thunder. The lowest musical note normally encountered is the deep shudder of the low C of large pipe organs. The upper limit of human hearing varies with the age of the listener. As a rule, only the young can hear frequencies above 20,000 Hz. When a man reaches the age of thirty or so, the upper limit of his hearing usually declines gradually to 15,000 Hz or even lower, especially in our noise-polluted environment; and with advancing age, hearing usually declines to 10,000 Hz or less.

Overtones and Tone Color

No musical instrument has a basic pitch higher than about 5000 Hz, so a system with a response limit of only 5000 Hz would still reproduce all the notes in music. But there is more to music than just pitch alone. There is the color, or timbre, which depends almost entirely on sounds much higher in frequency, mostly between 8000 and 15,000 Hz.

Suppose you play middle C on a cello, then play middle C on the trombone. Both are playing the same musical note, but you can easily tell them apart. Most people can detect even less distinctive differences in timbre. Two musicians playing the same instrument may elicit slightly different sounds from it because of their individual playing techniques. And the perceptive ear might even distinguish very subtle differences in tone between a Steinway grand and a Baldwin. What accounts for this?

When the great German physicist H. L. F. Helmholtz first began to analyze sound some one hundred years ago, he discovered that what the listener hears from an instrument as a single musical note actually consists of many different tones. There is first of all the basic pitch perceived by the ear, called the fundamental. In addition to this fundamental tone, the musical note embodies a whole series of additional tones, called overtones or harmonics. These are multiples of the fundamental frequency (i.e., twice, three times, or four or more times the frequency of

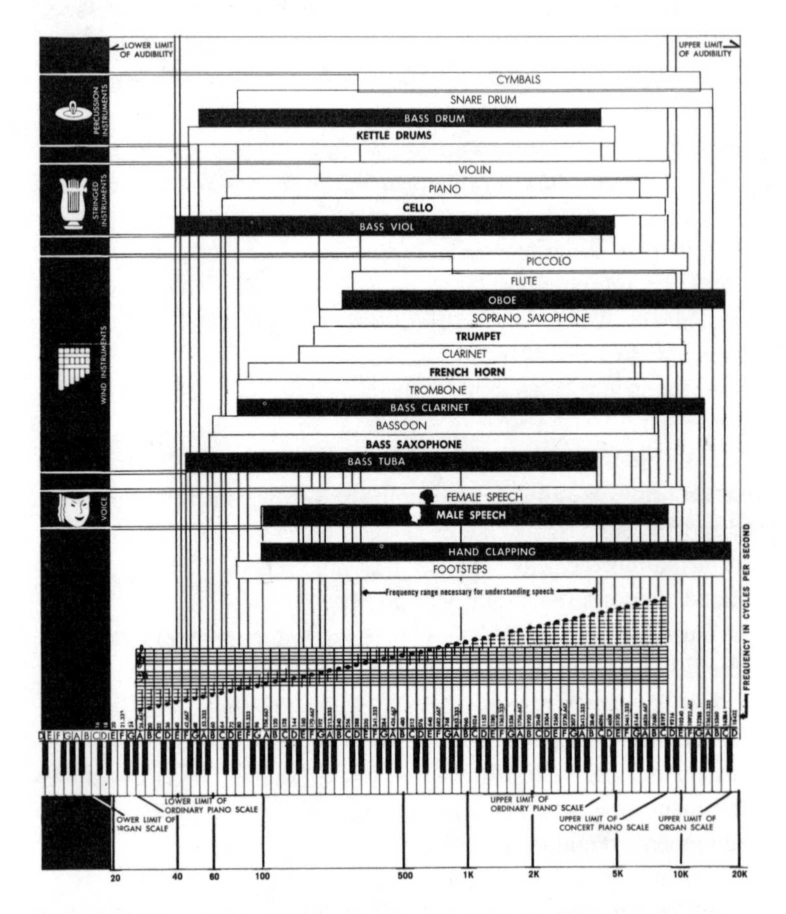

Frequency ranges of various musical instruments, the human voice, and other characteristic sounds. At the bottom of the graph, these ranges are correlated to the piano keyboard and standard musical notation. The lowest scale indicates actual frequencies in Hz. The abbreviation k stands for 1000. Thus 1 k = 1000 Hz.

the basic note). Not all of these overtones are equally strong. Each musical instrument has its own individual overtone pattern, and it is this pattern that gives each instrument the characteristic tone color by which we identify it. (If each musical note is actually a mixture of frequencies, so is each unmusical noise. But such sounds as water rushing into a bathtub or the wind rattling the shutters are made up of random frequencies that have no harmonic relation to each other. Hence they have no single, dominant frequency to be perceived as pitch.)

Hi-fi components must reproduce this whole pattern accurately if the sound is to be faithful to the original. This is why a wide—and uniform—range of frequency response is necessary. Suppose an oboe is playing a note with a basic frequency of 1500 Hz. Its overtones would be 3000 cps (1500 × 2), 4500 cps (1500 × 3), 6000 cps (1500 × 4), and so on. To do justice to the sound of the oboe, the system must reproduce these overtones in the exact relationships in which they occur in the original.

Even instruments such as the bass viol, the tuba, and the kettledrum—the lowest-pitched instruments in the orchestra—produce higher frequency overtones that give them their particular tonal flavors. This accounts for the seeming paradox that a sound system must have a frequency range up to at least 15,000 Hz in order to reproduce accurately instruments whose basic pitch is in the lowest octaves of the musical range.

The frequency response of a piece of sound equipment therefore indicates whether or not it is capable of reproducing a certain musical note truthfully. Since musical sounds generally fall within the range of 30 to 17,000 Hz, the response of good equipment should encompass this entire range.

"Flat" Frequency Response

Often you will find a frequency specification that reads 20–20,000 Hz, indicating that the range of response is even wider than required. The extra margin is all to the good, but such specifications in the form in which they are usually stated are not really very informative. All this really tells you is the highest and the lowest note the equipment can play. But from what we have explained so far about the nature of sound, it is clear that this alone is not enough to ensure fidelity.

What is needed additionally is *uniform* frequency response throughout this range. For it is the relative strength of overtones with respect to one another that makes an oboe sound like an oboe or a flute like a flute. To keep these overtones in correct proportion to each other, the sound system must reproduce the loudness of all tones in exact proportion to the original.

This is what is meant by the term "flat response" so often heard in describing components. A "flat" response means an even frequency response. It means that the equipment renders every sound from the lowest to the highest with precisely the emphasis musicians gave it: nothing is weakened, nothing unduly emphasized. This is an essential condition for natural sound reproduction. The term "flat response" derives from the engineer's method of plotting the frequency response on graph paper, as shown in the illustrations on page 53. These curves show the response of two hypothetical components. Both have an identical range: 20 to 20,000 Hz. Yet one would sound sweet and natural, the other shrill and blatant. What then is the difference?

Compare the curves. One is relatively flat without humps and wiggles. The other looks like the skyline of the Himalayas. The up-and-down variations in each curve show the deviations from flat response. They are measured in units called decibels (abbreviated as db), which indicate differences in loudness. It takes a difference of 10 db to make one tone seem "twice as loud" as another; 3 db is the smallest readily apparent loudness difference in music, though sharp-eared listeners may discern differences as small as 1 db.

The ideal, of course, is perfectly flat response curve. In practice, anything that varies from perfection by no more than that just-noticeable 3 db sounds quite good. By this standard, the

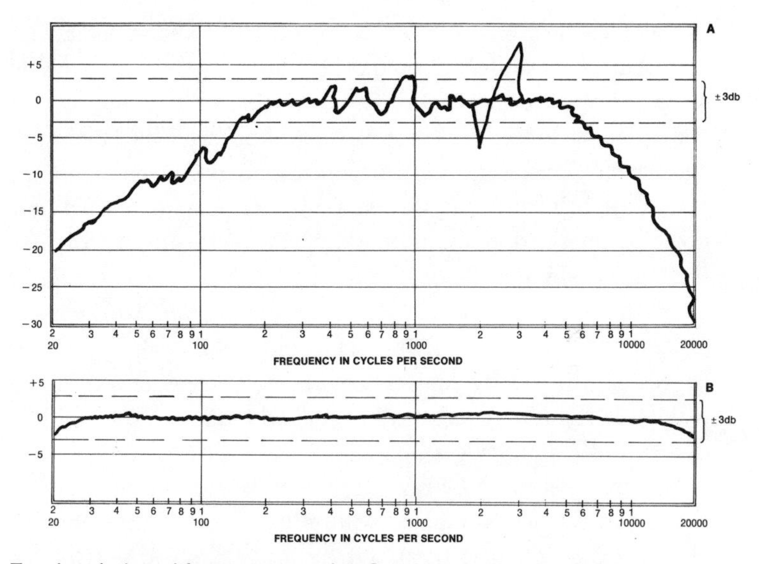

Two hand-plotted frequency graphs. Graph A shows a very uneven frequency response, falling off as much as 20 db in the bass and fizzing out in the highs above 10,000 Hz. Graph B shows a fairly "flat" response, remaining throughout the entire tonal range within 3 db of zero reference level.

illustrated component with the smooth curve is quite respectable, for we could say that its response was "20–20,000 Hz, ±2.5 db." This means that at no point in the entire tonal range does this component suppress or overstress any sound by more than 2.5 db, a deviation so small as to be virtually unnoticeable. The other graph shows deviations far greater, and both treble and bass response are dropping off by as much as 15 to 20 db. Such ragged response is characteristic of cheap sound equipment. All quality equipment should have response curves reasonably flat over the total musical range.

DISTORTION

This second of the three basic requirements for true sound is the vital factor that distinguishes first-rate equipment from lesser breeds.

Total absence of distortion is an ideal still unattainable. But in recent years designers of the best audio equipment have in some cases succeeded in keeping distortion so low that at normal listening levels it cannot be measured even by laboratory equipment. This is in marked contrast to the situation in the early days of high fidelity, when a fairly large amount of distortion was tolerated as the inevitable price one paid for extended frequency response. Today, with wide range response usually taken for granted, clean sound—that is, minimum distortion—is recognized as the new frontier in audio design. The great advances in the clarity and transparency of reproduced sound within the last few years must be credited to the far lower distortion of today's cartridges, amplifiers and speakers.

Extreme distortion betrays itself immediately through harshness and stridency, combined with gritty sound texture. But the blatant distortion that turns a trumpet into a kazoo (though common enough in the baleful blare of the usually inexpensive portable phonograph) is not the crux of the problem in high fidelity. Because distortion in high-fidelity equipment is fast approaching the vanishing point, the problem takes on a subtle and elusive aspect, both with respect to measurement methods and the subjective listening experience. Often the listener is not immediately aware of marginal distortion, but its presence has a subliminal, cumulative effect. So-called listening fatigue—a vague feeling of discomfort—is the ear's instinctive protest against the falsification of sound. Absence of listening fatigue, aside from adding to one's subjective pleasure, is a reliable indication of low distortion; for the observant ear is one of the most sensitive distortion detectors yet devised. That is why most audio manufacturers attempt to corroborate laboratory measurements with extensive listening tests.

Types of Distortion

Distortion occurs when the reproducing equipment alters the original sound—that is, when what comes out is no longer a true replica of what went in The two most common types of distortion are harmonic (HD) and intermodulation (IM).

Harmonic distortion derives its name not from the concept of musical harmony, but from the harmonics, or overtones, that are produced by an instrument in addition to its basic pitch. As we have said before, it is these natural overtones, or harmonics, that give the various instruments their individual tonal character. Harmonic distortion occurs when the amplifier (or any other component) generates harmonics of its own. These unwanted additions mix with the natural harmonics of the music and thus alter its tonal color.

To measure harmonic distortion in the laboratory a test signal consisting of a single frequency is fed into the amplifier under test. The output signal is than analyzed to see if the equipment has added any false overtones. The total amount of spurious harmonics is then measured and expressed as a percentage of the total output. As long as harmonic distortion remains below 1 percent of the total sound output, it has no perceptible effect.

Intermodulation (IM) distortion can occur when two or more different frequencies pass through the amplifier simultaneously, which is normally the case in playing music. The various frequencies and their harmonics then interact to generate new, discordant frequencies that were not part of the original sound. Mathematically, these new frequencies are the sums and the differences of the original signal frequencies. As an example, let us suppose that a cello is playing a 100-Hz note and a flute is playing a 1000-Hz note. If the amplifier has a bad case of IM, a 900-Hz tone and an 1100-Hz tone will be gratuitously added to the two original tones, these frequencies being the sum and the difference of the actually played notes. Moreover, the two false notes will interact, producing more notes, still further falsifying the sound. These intermodulation products add up to make the music sound harsh, shrill, raucous, and grainy.

To measure IM distortion in the laboratory, a low note and a high note are fed into an amplifier simultaneously (usually 60 Hz and 6000 Hz at a ratio of 4 to 1). At the output, the original test tones are filtered out, leaving only the intermodulation products. These are measured and again expressed as a percentage of the total sound output. If the IM distortion does not exceed 2 percent, an amplifier will yield respectable performance at its full rated power. Top-rated amplifiers in which IM distortion is less than 2 percent generally provide an added measure of clarity, which largely eliminates listening fatigue.

Figures for both harmonic distortion (sometimes abbreviated HD, or THD, for Total Harmonic Distortion) and for intermodulation distortion (IM) appear on the specification sheets of virtually all quality amplifiers, tuners, and receivers, and occasionally in tape recorder specifications as well. Generally, these figures should not exceed 1 percent.

But components are subject to other forms of sonic misbehavior that are rarely stated on data sheets because they are too elusive for direct measurement. One such culprit is called transient distortion. Transients are sounds that reach a high loudness level very quickly and then diminish just as quickly. Drum beats and other percussion sounds are good examples, as is the sound of plucked or struck instruments, such as guitars, cymbals, and pianos.

Transients present a problem to amplifiers and speakers because they represent pulses of voltage and current that build up suddenly to high levels. In the engineer's language, such sounds (and their electrical analogs in the amplifier) have a "steep wavefront." These steep wavefronts pose a stiff challenge, because the circuits within the amplifier must be capable of almost instantaneous—and controlled—response, and the speaker cone must respond without inertial delay. And when the brief sound pulse is past, the equipment must immediately return to the resting state without oscillation or bounce. Poor transient response distorts the steep vertical wavefronts, sounding the sharp edges, and sometimes causes what is known as "shock-excited oscillation." Aurally, such distortion is especially noticeable during percussive passages, but it can also affect the whole texture of the sound, beclouding orchestral transparency.

Unfortunately, there is no mathematically exact way of measuring the transient characteristics of amplifiers and speakers. Some amplifier manufacturers show photographs of oscilloscope square-wave patterns in their product specifications. A so-called square wave is used as a test signal because its vertical sides represent the steepest possible wavefront. Speaker manufacturers assess the transient response of their products by similar methods, feeding brief tone-bursts into the speaker. The speaker's output, picked up by a microphone and displayed on an oscilloscope screen, provides clues to the behavior of the cone when it is responding to a signal containing sudden starts and stops.

The trouble with these attempted measurements of transient distortion is that the waveforms encountered in music are somewhat more varied and complex than the fairly simple waveforms used as test signals. Consequently, no one-to-one relationship exists between test performance and musical performance. The data are indicative, not conclusive. In the end, the critical ear, listening for clarity of orchestral texture and equal coverage of the audible spectrum, remains the most useful test instrument in this elusive area.

To attain the goal of genuinely musical sound reproduction, all these different types of distortion must be kept within tolerably low levels, even at full volume. This relates to the available power of the amplifier, a topic treated in Chapter 5.

NOISE

Paradoxically, silence is one of the greatest virtues in a sound system. Yet since we do not hear silence, it becomes noticeable only by its absence. Let rumble from the turntable, hum from the amplifier, hiss from the tape, or atmospheric static with the radio signal intrude on the music, and we are quick enough to complain. But against a background of hushed stillness, we hear every musical detail with startling clarity. This adds to the music's emotional impact.

While we can't measure emotional impact, we can clearly specify in technical terms the amount of extraneous noise a sound system imposes on the music. As a general rule, the noise level of good sound equipment should be at least 40 db lower than the average level of music. This is still another yardstick by which to judge the performance capabilities of a system.

Our discussion here has centered on the physical measurement of the performance of a music reproducer. The question remains whether these technical factors provide a truly reliable index of quality for sound equipment. Granted, numbers cannot tell the whole story. Certain subtle nuances of sound elude physical description. They can be experienced only in subjective terms, by listening comparisons. Yet this should not lead anyone to dismiss engineering data as meaningless or uninformative. The listener who waves away all technical information and says, "Numbers mean nothing to me," deprives himself of the only objective standard of comparison in choosing his equipment.

It is of course entirely possible to enjoy good sound equipment without knowing or caring about its more technical aspects. After all, you don't have to be a cook to enjoy a good meal. Yet some understanding of gastronomy is likely to increase your appreciation of the finer aspects of culinary art. So it is with audio. The range of pleasure widens when technical awareness is added to musical perception.

4 Loudspeakers— The Voice of the System

A bachelor friend of mine once remarked that looking for a loudspeaker is like looking for a wife. The quest is for something to delight the senses, something that's easy to live with, preferably handsome, and a lasting joy. The trick is to pick wisely among a multitude of enticing possibilities.

A loudspeaker is basically a cone made of specially treated paper or plastic pushed back and forth by an electromagnetic device to re-create the sound vibrations of the music. The movements of the cone are the mechanical equivalent of the electrical signal fed to the loudspeaker. The moving cone imparts the corresponding vibrations to the surrounding air. In this way, the loudspeaker reconverts the electrical signal into audible sound.

More than any other component of your system, the speakers determine what you hear. They are, in a sense, musical instruments. Unlike the other components, which merely handle electrical signals, the speakers produce the audible sound. As the actual voice of the system, speakers represent a realm of engineering in which electronics, mechanics, acoustics, and musical instrument design overlap. With these diverse elements affecting loudspeaker performance, it is hardly surprising that virtually no two speaker designs sound quite alike. Each has a distinct "personality," its own kind of tone color.

This is a problem. Ideally a speaker should have *no* tone color of its own. It should merely reproduce the musical sound as is without adding its own flavor. But, unfortunately, such an ideal reproducer does not yet exist, though the best available speakers come remarkably close to this elusive perfection.

In picking a loudspeaker, you thus have a twofold aim: (1) You want to find a speaker with the least possible tonal coloration. (2) The inevitable residual coloration should be the kind that pleases your ears. In short, your personal taste in sound and the speaker's coloration should be compatible.

JUDGING SPEAKERS

Your individual judgment counts for more in speakers than in any other part of your audio system. *Yours* are the ears that should be satisfied with the speakers you buy. So let your ears be the final arbiters—not salesmen, friends, test labs, or whomever else you take as an authority.

On the other hand, if all of those authorities reject the par-

One of the first full-range loudspeakers was produced by Bell Telephone Laboratories in the early 1930s as part of their first experiments in high-fidelity sound. The famous conductor Leopold Stokowski was one of the guiding spirits behind these earliest efforts toward faithful music reproduction. Note the enormous woofer surmounted by two tweeters with subdivided horns to provide broader dispersion of highs. (*Photo: Bell Telephone Laboratories, Inc.*)
(2) Incredible as it may seem, this modern bookshelf speaker by KLH provides nearly the same frequency range as Bell Lab's historic monster, and with less distortion.

KLH

ticular speaker you have selected, it pays to find out just why you like what others don't. You may be responding to something bright and flashy in the speaker's sound, which may become annoyingly obtrusive in the long run. When in doubt, take a knowledgeable friend to the audio showroom with you and ask him to point out just why he disagrees with your choice. If you still can't hear what he's driving at, feel free to stick with your original choice. Chances are you'll wind up agreeing with him—and getting your ears educated at the same time.

My friend who picks a speaker as he would a wife happens to live within walking distance of Carnegie Hall. He goes to live concerts often and feels that the memory of the true sound of music helps him in judging speakers. "I go to concerts to calibrate my ears," he says.

I tend to believe that listening to live concerts is not very valuable as a standard of comparison in judging sound equipment. For one thing, human beings have a poor memory for sound, and I doubt that you can remember exactly your sonic impressions from the concert hall when you go to your audio dealer. Besides, listening in a living room is so different from listening in the concert hall that direct comparisons are not meaningful. What we're after is to create in the home the aesthetic equivalent of the musical performance, not necessarily the physical analog.

A record can be a better guide to speaker shopping. But it must be a record that covers a fairly wide range of frequencies and musical effects. And it should contain the kind of music you most often listen to. Pick a favorite you're particularly familiar with but make sure it is a new copy, not a worn-out disc. Play this record on every speaker you're considering, and you'll have a constant factor for comparison. Of course, you can try other records, too. But the same record played on all the speakers you're comparing will provide the essential reference.

Some dealers show off their equipment by means of special demonstration records. Stay clear of those. They usually feature spectacular sounds of brass, percussion, and organ, which are deceiving in themselves and often are recorded in a special way to make a poor speaker sound better than it is. Play the kind of music you normally listen to for your own enjoyment.

Symphonic music is the best test for a speaker because it features the full range of frequencies and the many different instruments provide a variety of sound textures. As you listen, be on the alert for these telltale clues:

1. Do the violins have a silky sheen of sound without harshness? This shows that the speaker has smooth treble response.
2. Is there solid weight in the sound of cello and contrabass? This is a test for bass projection.
3. Do drums and other percussion instruments or plucked strings sound sharp and crisp? This is a test for so-called transient response.
4. Does the sound stay clear and unblurred even in loud passages scored for full orchestra? This test shows up distortion.

Another important rule for comparing speakers: Make sure you're hearing all the speakers under test at exactly the same loudness level. Even a very slight difference in loudness can fool the ear and make the louder speaker seem the better, regardless of other factors. Also, compare only two speakers at a time. Otherwise you can't possibly sort out your impressions. After finishing the test of one pair, compare the better of the two with another, and so on. The main thing is to make sure that all the speakers you are comparing sound equally loud at the time of comparison.

Clarity of texture is probably the most important single attribute of a good speaker. When a full orchestra plays, can you discern the individual instruments? Or do you just hear a glob of homogenized sound? Clarity indicates that the speaker adequately covers the upper range of the sound spectrum. It also shows that its frequency response is reasonably flat, that it doesn't overstress or underplay any part of the range.

The bass tones should have solidity, weight, and—where the

music calls for it—power. But where there's little or no bass in the music, little or none should be heard from the speaker. Some poorly designed speakers give an impression of more bass than they actually have, because they add false bass to almost everything they play. The bass in such speakers thumps along at a

A modern loudspeaker, with its decorative grille removed, reveals its innards. The large woofer at the bottom is augmented by two separate mid-range units, while the treble is projected by the tweeter at top center. The enclosure is fully sealed in accordance with the acoustic suspension principle.

constant note regardless of the music's actual pitch. You can hear this, with all its hollow boom, from any jukebox. Tubby bass like this is caused by unwanted bass resonances that falsify the true character of sound and blur the clarity of the music. Another good test is to tune in an FM station and listen to the announcer's voice. If the speaker has such false resonance, the announcer will sound boomy, as if speaking inside a barrel. An accurate speaker will make his voice seem as natural as if he were speaking with you in the same room.

Some speaker designs, usually the smaller and cheaper ones, try to give the impression of stronger bass by so-called doubling. Doubling is a kind of distortion in which the speaker reinforces the bass by sounding a spurious note one octave higher than the true note of the music. The best way to spot this type of distortion is with a test record that features a descending scale of "pure" tones in the bass. The onset of doubling will then become immediately apparent as a change in tone color. The best speakers in any price class are free from these attempts to magnify the sound.

High frequencies should be smooth and clear sounding, not harsh or gritty. Massed violins should blend smoothly, yet the initial "bite" of bow on string should be clearly audible on solo violins. Walk around as you listen, to be sure that the highs are evenly dispersed throughout the listening area; the sound should not become noticeably brighter as you stand directly in front of the speaker, nor should the highs disappear as you move to either side.

The noise you hear between FM stations on a tuner (see Chapter 10) is an excellent test signal; if a speaker's treble dispersion is faulty, it will show up as a sharp change of the sound's character as you move from front to side. After you've heard FM interstation noise through several speakers, you'll begin to recognize it as a test for sonic balance too—a nasal quality, a sharp hissiness, or a definite sense of pitch anywhere within the frequency range are all signs of speaker defects.

SPEAKER SPECIFICATIONS

Published specifications about the ability of a speaker to meet these requirements are ambiguous. Comparing specs, you may find a $40 speaker and a $500 speaker claiming the same frequency response. Does this mean they are equally good? That seems unlikely, considering the price difference. So you ask, "Are the figures lying? Or are the liars figuring?" The answer is: a bit of both. Because engineers, just like accountants, find ways of making the figures look better than the facts. The specs may be true as stated, but that doesn't mean that the $40 speaker sounds like the $500 model. Suppose the specs say that a speaker has a frequency response from 30 to 18,000 cycles. All this tells you is the top and bottom notes the speaker can squeeze out. It tells you nothing about the quality of sound. The

Another common type of loudspeaker design is the so-called coaxial speaker, in which the tweeter, equipped with multiple projection horns, nestles inside the large cone of the woofer. (*Photo: Altec Corp.*)

bottom bass, for instance, might just be a hoarse rattle instead of a deep rich tone. And that top treble might screech like an old-time street car on a curve.

What really counts is the speaker's ability to handle the range between its upper and lower response limits with good dispersion, low distortion, and a fair degree of uniformity. And when a speaker manufacturer is really serious about presenting this information, he does it not with figures, but with graphs.

Ideally, the speaker should give every note its exact due, making it neither louder nor softer in relation to all the others than it was in the original performance. When this kind of frequency response is plotted on paper (loudness against pitch) the resulting graph is a straight, flat line. That is why engineers speak of flat frequency response as the theoretical optimum.

No actual loudspeaker, alas, attains such flat response. Like all physical objects, the speaker cone is beset by a variety of problems, including natural resonances that cause it to emphasize some frequencies and deemphasize others. The net result of such idiosyncrasies is a change in tone color. To some degree, every speaker adds its own tonal coloration to the music; the job of the designer is to keep such individuality subdued so that the speaker's own tone color doesn't interfere with that of the music.

One laboratory procedure for testing a speaker's frequency response is to place a calibrated microphone in front of it and play a test signal that sweeps through the whole range of audible frequencies. To keep room resonances from affecting the measurement, these tests are frequently carried out in anechoic chambers, special rooms designed to eliminate all sound reflection. To obtain the speaker's off-axis frequency response (important in regard to "openness" of the reproduction) either several microphone positions are used, several microphones are used simultaneously, or the speaker is rotated. These curves may be averaged together or superimposed.

Frequency-response graphs obtained in this manner are usually far from flat. Most of them look like Alpine landscapes, full of ragged peaks and precipitous crevasses. Only long experience in testing many different speakers teaches how to relate the profile of such a graph to the actual sound. However, only deviations greater than about 5 db extending over a broad span of frequencies are apparent to the ear. For example, elevations of the frequency curve in the region of 7000 to 12,000 Hz may tend to make a speaker shrill and harsh. A similar hump at 100 to 200 Hz might make it boomy or tubby. In any case, for subtler discriminations, the evidence of the graph is never conclusive, and it is the ear rather than the test instrument that must render the final judgment.

Don't be dismayed by the jagged appearance of speaker response curves, especially as compared to the far smoother curves of other components. While vagaries of frequency response apply in some degree to all audio components, they are far harder to control in speakers than in electronic circuits. For, like any vibrating object, the loudspeaker has natural resonances of its own that obtrude stubbornly. To smooth out these resonances, speaker designers use various techniques, usually having to do with the choice and treatment of their cone materials, and they tend to be as closemouthed about them as a fiddlemaker is about his woods and lacquers. If a loudspeaker manufacturer publishes a response curve at all, chances are he has reason to be proud of it. Too smooth a speaker curve, in fact, would be suspicious. Such curves have almost certainly been smoothed out by an advertising artist's hand, not by the speaker designer's skill.

While frequency-response graphs often include some information on treble dispersion, some manufacturers plot dispersion separately, for several frequencies. These curves may be plotted on a full circle, or just over a significant section thereof as in the illustration on page 80. And polar response curves for different frequencies may be graphed separately, as they are here, or superimposed.

Since no speaker has flat frequency response, all speakers

have some tonal coloration. Once you have found the least inaccurately colored speakers in your price range, you'll be able to balance the types of coloration—tonal warmth in one speaker as against brilliance in another—and select the speaker that will please you best.

But frequency range and dispersion, however important, are not the only indices of speaker performance. Much of the clarity of a loudspeaker depends on its transient response. Transient response is a measure of the speaker's ability to keep in step with the electrical impulses fed to it by the amplifier. In the hotel business, a transient is a guest who stays only a short time. In audio, the term has a similar meaning. It refers to sounds that start fast and are gone soon, like drumbeats, cymbal crashes, the clock of woodblocks, the plucking of strings, or the clang of bells at the moment the hammer strikes. Poor speakers cannot handle these sudden sound bursts. Their cones have too much inertia. They are slow to respond to the sudden sound impulse and they keep on jiggling long after the sound has stopped. Result: the music sounds blurred and soggy; the sharpness and excitement are missing. In a speaker with good transient response, drums, other percussion instruments, and plucked strings will sound sharp and crisp, and you can easily sense the moment of percussive contact between drumstick and drum or between hammer and piano string.

It is hard to believe that such brief sonic episodes can be so important; yet it is precisely these tiny elusive elements of sound that make up the texture of an orchestral fabric and convey much of the music's emotional impact. Unless the speaker reacts instantly to these sudden explosions of sound, the aural impression is blurred and its emotional force weakened.

Even the subtler aspects of music depend greatly on good transient reproduction. Take the plucking of strings, the accents produced by the tonguing of reed instruments (including certain stops of the organ). The sum of such sonic detail makes up the distinctive sound texture of a score. A speaker failing in this respect falsifies and obscures an element of music that is far more important to artistic communication than is generally realized.

To assess a speaker's transient response, some manufacturers set up tone-burst tests consisting of very brief test tones in various frequency ranges fed to the speaker from a special test oscillator. A high-quality microphone positioned directly in front of the speaker monitors the speaker's reaction to these tone bursts, displaying a visual trace of speaker response on an oscilloscope screen. Some manufacturers publish photographs of these "scope traces" as part of their speaker specs.

Unfortunately these traces can be relatively meaningless, if not actually misleading, unless evaluated in the perspective of other data and long personal experience. The tone bursts are merely a simplifying assumption for analytic purposes and do not truly represent the many complexities of interwoven musical sound patterns. The best to be said for these kinds of tests is that they help to identify the more obvious shortcomings of a speaker, but a faultless scope trace is in itself no guarantee of musical excellence. The best testing device for "clean" transient response is the musically perceptive ear, critically intent upon discerning crispness of percussive attack and clarity of orchestral texture.

Speaker designers have various ways of controlling transient response, and they sometimes allude to them in catalogs of specification sheets. Good transient response requires that each movement of the speaker cone be so controlled that it does not overshoot its mark. Also, the cone movement must respond immediately and without sluggishness to the electric impulses representing the music. This requires a strong magnet that can firmly control the motions of the cone. A light, flexibly suspended cone also helps because it gets moving faster and is easier to control during sudden starts and stops.

But spec-sheet references to details of magnet strength and cone or rim construction must be taken as relating only vaguely to the actual transient response of the speaker described. References to magnet weight and material will tell you little. The more significant measurement is magnetic flux density. It is expressed as a number of "lines of flux," and it measures the magnetic force itself. The higher the number, the more accurately cone motion keeps in step with the electric sound signal. But different speaker designs may require more or less flux to attain the same degree of control over cone motion. So this, too, is not as good an indicator of quality than your own ears.

Because they are mechanical rather than electronic devices, loudspeakers have higher distortion than most amplifiers and tuners. Now, however, technological developments have reduced speaker distortion to the point where manufacturers are a little less reluctant than they used to be to publish distortion figures. If a speaker manufacturer cites any distortion figures for a speaker, it is probably a superior unit. Some manufacturers supply not only distortion figures but distortion curves, which are even more revealing.

Among speaker specifications you will also find such items as power requirements, power-handling capacity, and impedance. These require a brief explanation.

The power requirement tells you how many watts your amplifier must deliver on each channel to drive the speaker adequately. The power-handling capacity tells you the maximum power the speaker can absorb without overloading. Both these figures are no more than rough approximations. For all practical purposes you only have to make sure that you pick an amplifier with enough power per channel to drive the particular speaker you choose. This means that the wattage per channel should be equal to or higher than the power requirement of the speaker.

Impedance is a measurement necessary for matching the amplifier to the speaker. Most speakers are designed for 8 ohms' impedance, though occasionally you may find a model with 4 or 16 ohms' impedance. As long as your amplifier has output connections at the same impedance as that of the speaker, you will have no trouble matching them. Since virtually all amplifiers have output terminals for 4 and 8 ohms, most amplifiers can be matched to most speakers as long as the power requirement is satisfied.

Cone design and material can affect distortion, too, especially in the higher frequencies. Trying to handle many frequencies at once a speaker cone's vibration may "break up"; different portions of the cone then vibrate at different frequencies, interfering with each other to create distortion.

SPEAKER DESIGN

The most common type of speaker is driven by a coil, which is connected to the amplifier. The coil is surrounded by a magnet and moves back and forth as the magnet's field interacts with the varying magnetic field radiated by the coil as the amplifier's varying signals rush through it. The coil, in turn, moves a cone, which pushes and pulls the air to make sound waves. The rim, or suspension, of the speaker is a flexible ring holding the speaker cone at its edge. Frequently, a speaker system will contain more than one speaker. And the more expensive the system, the more speakers it is likely to include. Some manufacturers—chiefly the makers of consoles and cheap phonographs—tout their packages as having six or more speakers, slyly implying that more speakers automatically mean better sound—which is not necessarily so. Six poor speakers in a box will not sound better than two good ones.

The best low-cost speakers usually use only one speaker unit, in the enclosure, most often an eight-inch type. That's big enough to provide decent bass and small enough to disperse its treble frequencies reasonably well throughout the room. The very lowest and very highest frequencies are weak or missing, but what is left can be quite well balanced, and eminently listenable.

A multi-speaker system *can* cover a wider frequency range

more evenly, with lower distortion and greater dispersion to boot. This is because low and high frequencies impose different requirements on loudspeakers and should be reproduced by separate speaker units.

Woofers and Tweeters

Bass frequencies are handled best by woofers, large, rugged speakers designed to move large quantities of air. Most woofers have heavy cones and highly flexible (so-called high-compliance) suspensions, to keep their resonant frequencies low and to let the cone push out heavy bass thrusts without distortion.

Tweeters, which reproduce the high frequencies, have tiny cones of stiff, lightweight materials that can vibrate rapidly without buckling. Small size helps them achieve better dispersion too. Dispersion can also be increased by using two or more tweeters angled away from one another by making the tweeter diaphragm a convex dome rather than a concave cone, and through the use of such devices as acoustic-lens deflectors or flared metal horns.

Some systems also have separate mid-range units to tackle the middle frequencies from about 800 to 10,000 Hz. Others add super-tweeters for maximum dispersion of the very highest frequencies. In each case, this division of labor lets the designer give each of his speakers the best characteristics for the frequency range it handles, without the compromises inherent in single, full-range speakers.

With each increase in complexity, from single speaker to two-way (woofer-tweeter) system, all the way up to four-way systems with woofers, mid-range speakers, tweeters, and super-tweeters, the design problems increase. Expensive crossover networks must be used to feed each speaker its proper range of frequencies; care must be taken that all speakers in the system are in balance. All speakers must begin to move at once in response to a signal to avoid time-delay distortion. These problems are all soluble, but solving them raises the system's cost. This is why inexpensive multi-speaker systems may sound worse than single-speaker systems selling at the same price.

For that particular price, one manufacturer might produce an excellent single-unit speaker while another manufacturer tries to produce a multi-unit speaker with separate woofers, tweeters, and mid-range units which—even in aggregate—may not match the quality of a well-designed but less complex speaker. On the whole, a simple woofer-tweeter combination usually represents the optimum solution between such factors as cost, complexity, and attainable performance.

Speaker Enclosures

The same diversity of valid approaches applies to speaker enclosures. The enclosure, sometimes called a baffle, is far from just a box for the speakers. It plays a vital part in sound reproduction. Without an enclosure even the best speaker would sound thin and reedy; virtually all the low notes would be missing. The reason for this is that both the front and back of a speaker cone emit sound waves. On its forward trip, the cone compresses the air in front while creating a partial vacuum in back. On the backward trip, the reverse occurs. At all times the front and the back waves are out of phase; the air is compressed on one side of the speaker and rarefied on the other. Consequently, if the front wave meets the back wave, the two just cancel each other, and no sound is produced. This effect is especially noticeable at low frequencies.

The main job of the enclosure, therefore, is to prevent a collision between the sound waves from the front of the speaker cone and the waves from the back. There are a number of ways in which enclosures accomplish this. One of the simplest is to contain the back wave in the box by sealing the enclosure airtight and lining it with sound-absorbent material. This is called an infinite baffle, because, in theory, it completely isolates the back wave as a space of infinite dimensions would. Such an enclosure must be extremely sturdy and firmly braced to keep its panels from vibrating and radiating sound waves through the enclosure back into the listening room.

In the early days of high fidelity, infinite baffles were usually quite large so that when the air in the box was compressed by the backward motion of the speaker cone it did not hinder the cone's movement. A more recent variant of the infinite-baffle principle, the acoustic-suspension speaker, is designed to use the back pressure of the confined air in the enclosure. This system calls for a speaker whose cone is so loosely suspended that it requires the back pressure of the confined air to function properly. Because a relatively small volume of air will supply the necessary pressure (larger volumes of air being too elastic), the acoustic-suspension loudspeaker actually *requires* an enclosure of fairly small proportions. Thus compactness becomes a necessity rather than a compromise.

Since the back wave in a sealed enclosure remains entirely shut up in the box, half the sonic energy produced by the speaker never reaches the listener's ears. As a consequence, these acoustic-suspension speakers tend to be somewhat inefficient. This is not a disparagement of their quality; it merely means that a relatively high amount of amplifier power is required to produce a given loudness. There are, however, alternative enclosure designs that allow the back wave to contribute part of its energy to the audible output of the speaker. Consequently, such enclosures are less demanding in terms of amplifier wattage.

For amplifiers of moderate power, of less than 20 watts per channel, it is preferable to use loudspeaker enclosures that let the back wave contribute to the audible output of the speaker and help the speaker attain adequate loudness with less wattage. The most popular enclosures of this kind operate on the bass-reflex principle; the back wave is let out of the box through a carefully calculated opening known as the port or vent. In some designs the vent is joined to an interior duct, which reduces the required enclosure size.

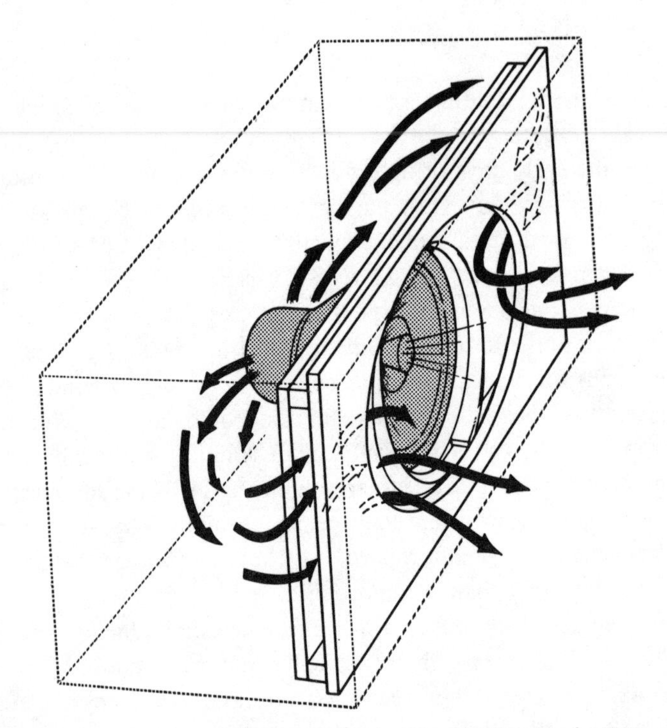

The sound-path in a so-called vented enclosure permits sound waves from the rear of the speaker to emerge from an opening in front. This yields more sound per watt (i.e., higher efficiency) but usually increases unwanted resonance.

The liberated back wave does not cancel the front wave of the speaker, because the path it follows in the enclosure is just long enough to reverse its phase by the time it emerges from the port. It is now in phase with the speaker's direct radiation and reinforces instead of canceling it. Offhand, this would seem an ideal solution to the problem of taming the troublesome back wave and, at the same time, efficiently converting amplifier wattage into acoustic energy. However, there is one hitch. Any cavity or vented enclosure has its own natural resonance, whose pitch depends on the air volume it contains. If the music to be reproduced has bass notes near the enclosure's natural resonance, this resonance will be added to the sound, falsifying its character and making it boomy. If you have ever tried talking into a barrel, you know the effect. Speaker manufacturers get around this by matching the resonance of the enclosure to the resonance of the speaker cone in such a way that the two resonances largely suppress one another. Consequently, well-designed speakers of this kind are free of noticeable boominess. And by interior padding of the box and various special treatments of the vent, it is possible to keep coloration at a minimum.

The most efficient of all loudspeaker enclosures, requiring very little amplifier power, are the horn systems, which work on still another principle. They use a tapered duct, acting like a megaphone, to transmit the sound from one side of the speaker cone into the listening room. Inherently nonresonant, horns present no problem of boominess; but to be effective over the whole audible range, they have to be quite long and their openings must be large. To make enclosures of this type presentable in the living room, horns are usually doubled up and folded on themselves to conserve space, and the whole structure is housed in attractive furniture, usually designed for corner placement. Still, such enclosures tend to be fairly big. But their owners usually claim that the richness of sound obtainable from horn enclosures justifies their space requirements.

These are by no means all the possible enclosure types, merely the best known. Of these, the acoustic-suspension type is by far the most popular today and can be most generally recommended for practicality, compactness, and sonic merit. The compactness of these speakers makes them especially suitable for urban apartments where living space is limited. Besides, most of today's best sounding speaker systems are of the acoustic-suspension type.

It is essential that the characteristics of any enclosure complement those of the speaker it contains. If you are buying speaker systems complete with their own enclosures, you need not worry about this. But if you buy speakers separately, follow the manufacturer's recommendations for a suitable enclosure.

BIG VS "BOOKSHELF" SPEAKERS

The acoustic-suspension speaker has also helped dispel the notion that only a big speaker can produce a "big" sound and only a big woofer can produce low bass. "Big" sound owes its impressive quality to a combination of a broad treble-dispersion (which has nothing to do with the enclosure but depends on the tweeter design) and a solid bass. The bass depends on the amount of air the speaker can push. The more air that is moved at low frequencies, the deeper, richer, and more powerful the low notes appear.

There are two ways of moving the necessary volume of air. You can use a very big woofer—fifteen inches or more in diameter—that does not move back and forth very much. Until the invention of air-suspension speakers, these big woofers were the only way to produce good bass, thus giving rise to the myth about the necessity of big speakers. But, in fact, just as much air can be moved with a smaller woofer that has a wider stroke —i.e., one that moves a greater distance on its back-and-forth swings. With its longer swings the small speaker can pump as much air as a big speaker with shorter swings. That is exactly what acoustic-suspension speakers do, which is why a good "bookshelf" speaker can provide as much bass as some large floor models.

MULTIDIRECTIONAL SPEAKERS

A special breed of speakers you may encounter are the so-called multidirectional models. Unlike standard loudspeakers, which radiate their sound mainly in a forward direction toward the listener, these "omnis" scatter their sound in several directions. Some of them aim their sound toward both front and rear, others have additional sound projectors toward the sides, while still others are omnidirectional, scattering sound in a full circle.

When sound is scattered in this manner, most of it is aimed away from the listener and reaches him only by reflection from the walls of the room. As one engineer put it, "It's like a carom shot bouncing off the edge of the pool table before hitting the pocket."

The elements of a floor-standing loudspeaker are laid out in front of it: two woofers, one midrange unit, and a tweeter.

Proponents of the idea claim that this type of sound dispersion more closely simulates a live music listening situation. A listener sitting at a concert at some distance from the stage hears only about 20 percent of the total sound directly from the source. Most of his sound impressions reach him by way of multiple reflections from walls, ceiling, and floor. The exact ratio between direct and reflected sound he hears varies according to seat location and hall acoustics. But in most situations reflected sound predominates and is the essential element in determining his acoustic impressions. Multidirectional speakers are an attempt to create analogous sound output away from the listener.

They encourage wall reflections in the home similar to those in a concert hall.

The ratio of direct to reflected sound is just one of several factors accounting for the effect of multidirectional speakers. Another is the multiplicity of sound pathways. Because the different reflection angles create different path lengths between sound source and listener, each path represents a different arrival time of the sound at the listener's ears. This may be only a small fraction of a second, but in aggregate these differences contribute to the acoustic impression made by these speakers. Subjectively, most listeners notice two distinct effects: (1) an increased sense of depth and spaciousness of sound; and (2) a wider spread of the area in which the stereo effect is noticed, thus making the listening position less critical. Asked to describe their impressions, listeners typically use phrases like "being immersed in a sea of sound," or "it sounds as if the room were twice as big as before."

One explanation for the apparent increase in room size is the creation of mirror images of sound at the reflection points. At the higher frequencies, which contribute most of the multidirec-

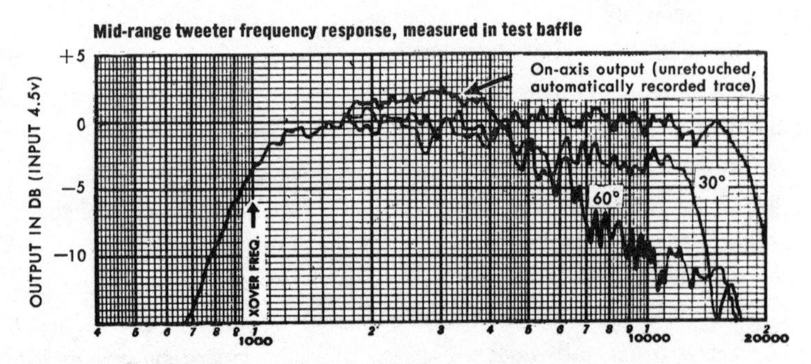

Directionality of a loudspeaker is shown in these response curves all taken from the same midrange speaker. One set of measurements was taken directly in front of the speaker ("on axis"). The other sets were taken at angles of 30° and 60° respectively. Note that upper-range projection falls off toward the side, indicating uneven dispersion of highs. (*Chart: Acoustic Research, Inc.*)

tional effect, acoustic reflection is analogous to light reflection in a mirror. In optical mirrors, the image does not appear to be directly in the plane of the mirror but behind the mirror. Similarly, in acoustic reflection, the reflected sound seems to come from *behind* the wall. To the ear, it then seems as if the walls were pushed farther back and that the sonic panorama extends beyond the confines of the room. This sort of *trompe d'oreil* is especially impressive when orchestral music is played in a relatively small room, for it creates the impression of a broad sound stage even within a closely confined environment.

By no means everyone is convinced of the merits of the multidirectional approach. Opponents concede that there might have been merit in the multidirectional principle some years ago when most conventional speakers radiated treble in a narrow beam. Today, because speakers with broad dispersion are so common, these engineers feel that the omnidirectional approach has become a solution in search of a problem.

From a purely practical viewpoint, the multidirectionals have a certain handicap. To radiate in several directions, they must be free standing, and they cannot be placed in bookshelf enclosures. Another objection is that by broadening the apparent sound source, the "multis" sometimes create curious exaggerations in reproducing soloists and singers, such as sopranos who seem to be eight feet wide. How much of this kind of space distortion there is depends on the acoustics of the listening room interacting with the speakers.

SPEAKER PRICES

With the price range of good speakers ranging all the way from about $50 to $200 or more, the question naturally arises: How much should you pay for a speaker? For a budget system, several fine models are available for about $50 to $80 per speaker from such renowned manufacturers as Acoustic Research, Advent, KLH, Dynaco, and Electro-Voice. For a deluxe system, you may pick speakers in the $200-plus price bracket. There is a saying among audio experts that half of your total sound budget should go into speakers, the other half into the rest of the equipment (not counting tape recorders). Supposedly this breakdown gives you a pretty good quality match between your speakers and the other gear. There is some merit to this argument, but modern speaker design has lowered the cost of quality. Today you can find loudspeakers in the $100-$150 price range whose sound is as good as any, regardless of price. The choice, ultimately, is a personal one. In this chapter we have discussed some of the technical quality standards for speakers. As long as these are met, the remaining differences are purely subjective. It is like choosing between a Steinway piano and a Baldwin, between one fine violin and another. No engineering formula can describe just what it is that gives the greatest pleasure to your ears.

5
Amplifiers I— Power Politics

The main job of the amplifier is to take the weak electric signals from the record player, the radio tuner, or the tape recorder and enlarge—or amplify—them so that they become sufficiently powerful to drive the speakers.

This suggests the first question to ask yourself when picking an amplifier. How much power do you really need? Of course, it's easy to say the more the better. But the price of amplifiers goes up rather sharply with their power rating, and there's no point in buying extra power you'll never really use. On the other hand, if your amplifier is too puny to deliver full, rich sound in the particular setting of your home your musical enjoyment will be curtailed. A rational purchase therefore depends on formulating the right kind of "power politics" for your particular needs. What then are the relevant factors?

THE MEANING OF POWER

First let's clear the air of popular misconceptions. A lot of watts doesn't necessarily mean a lot of sound. A 100-watt amplifier, for example, doesn't play ten times as loud as a 10-watt amplifier since the human ear does not translate power output into a proportional sense of loudness. The audible difference in loudness between an amplifier producing 25 watts output and one producing 50 watts is only 3 db, a comparatively small increase. Why, then, pay a premium for those extra watts if you can hardly hear them? The answer is that mere loudness has little to do with fidelity in reproducing music. What you gain from the added wattage is not added volume, but clarity of sound in loud passages.

Amplifier power can be compared to horsepower in cars. You do not always drive your car with the gas pedal pressed down to the floor, extracting every bit of available power from the engine. Similarly, the amplifier rarely operates at full output. But there are moments in music—just as there are moments on the road—when ample power reserve helps you over a steep hill. In music, these "hills" are orchestral climaxes, crashing fortissimi, full chords struck forcefully on the piano, or the deep rolling tones of the double bass or the organ pedals. Such sounds represent tremendous concentrations of acoustical energy. To maintain clarity of sound in these passages an extra margin of amplifier power is needed.

Suppose you have a rather small amplifier that delivers about 12 watts per channel. Playing music at normal room volume with moderately efficient speakers would not overtax such an amplifier during the quieter passages. However, when the score calls for drums and trumpets fortissimo, the sound would be loud, but because of the insufficient power reserve available, the amplifier would momentarily veer into distortion. Without an adequate power reserve, the amplifier would "clip" the tops and bottoms of the waveforms that exceed its 12-watt rating. The clipping represents severe harmonic and intermodulation distortion. This often happens within a fraction of a second—during the instant the piano hammer hits the strings or the stick crashes against the drumhead—so fast that the listener may not be aware that he is hearing distortion. Yet the musical climax becomes clouded, losing its immediacy, and the result over any extended period is listening fatigue. In contrast, an amplifier with a sufficient reserve of power glides smoothly over such tonal hurdles, thereby allowing crucial passages to come through undistorted and reproducing faithfully the sound of the music as heard during the recording session. In short, the difference between large and small amplifiers is not so much in overall loudness but rather in the quality of sound at volume peaks.

Subjectively, then, inadequate power and the distortion it produces during sonic peaks give the listener a vaguely uncomfortable feeling about the overall sound texture. This is one of the main factors in "listening fatigue"—the odd feeling of discomfort and irritation the musically sensitive listener experiences in the presence of distorted sound. By contrast, with an amplifier that has sufficient power reserve even heavily scored passages of orchestral music remain clear and transparent, and you can listen happily for hours without getting that edgy feeling.

Higher output power also helps pump out better bass. To get an idea of the energy contained in the really low notes, visualize such instruments as kettledrum, tuba, and bass fiddle. Think of the sheer physical force required to play them. To reproduce this energy in your living room takes a lot of extra watts. And when it comes to bass reproduction, the official wattage ratings can be misleading. Suppose your amplifier is rated at 40 watts. Ideally, it should then put out a maximum of 40 clean, undistorted watts all the way from top to bottom of the tonal range. However, power measurements on many amplifiers are made in the mid-range, around 1000 Hz. When the music gets down to the very bottom—like a thumping wallop in the bass around

35 Hz—that amplifier may not be able to squeeze out more than 5 watts without splattering distortion all over the place. The result is that the music lacks a feeling of power and solidity in the low range, where it's most needed. Besides, because distortion is introduced by the power surge in the bass, the overall sound texture becomes muddy.

Again the solution to this problem is plenty of spare power. An amplifier with a power rating of, say, 80 watts has a lot more power left over at the low end than an amplifier of 40 watts, even though the actual power measurement is made in mid-range and full power may not extend all the way down to the bass. To give you a clear idea how an amplifier performs at extreme lows or highs (not just in the mid-range) some manufacturers of high-quality gear include a so-called power bandwidth in their amplifier specifications. This tells you the range of frequencies that the amplifier can deliver at at least half its rated power. Don't confuse it with the regular frequency response curve, which tells you the amplifier's frequency limits at a much lower power level, usually at 1 watt output. This represents the amplifier's performance under less stringent conditions, while the power bandwidth represents its performance at critical peaks.

HOW MUCH POWER IS ENOUGH?

Having clarified the general importance of adequate power, let us return to our original question: How much power do *you* need? The answer depends on three factors: (1) your choice of speakers, (2) the size and furnishings of your room, and (3) your musical tastes. Let us consider these one by one.

Some speakers are more efficient than others. This simply means that they'll give you more sound per watt of amplifier power. However, this efficiency is not the only thing that counts, and to say that a speaker is inefficient is not to disparage it. In fact, today's best loudspeakers—especially the popular, compact bookshelf models—are notably inefficient. They may sound extremely good, but they gobble up a lot of watts to produce a room-filling loudness level. Most modern speakers operate on the so-called acoustic-suspension principle, which makes it possible to project low bass from a comparatively small enclosure. Today's bookshelf speakers produce as much bass as the room-filling monsters of yesteryear. But they can do it only because their designers have traded off size against efficiency. The speakers are more compact, but it takes more power to drive them.

So your first guideline in determining the amplifier power you need is to look at the power requirement of the speakers you have chosen. Let's say the minimum power requirement of your speakers (according to their manufacturer) is 15 watts each. This figure is fairly representative of modern acoustic-suspension speakers (which some manufacturers may call air-suspension speakers). In a normal room 15 watts should allow a comfortable power margin even for full orchestral passages.

But what is a normal room? For the purpose of this discussion, we assume that it has a volume of anywhere from 2000 to 3000 cubic feet. If your room is bigger, the basic power requirement will also be greater. For 4000 cubic feet you may need as much as 20 watts per channel, and for 6000 cubic feet you will need 30 watts. If your home boasts a baronial 8000-cubic-foot living room, triple the basic power requirement to 45 watts per channel. Since all the wattage figures here are given on a per-channel basis, the total power output of your amplifier should be twice the figures stated above.

That's only the beginning of our calculation. Next, enter your furnishings into the equation. Suppose most of the surfaces in your home reflect sound rather than absorb it, with smooth plaster walls, hard floors with just a few scattered rugs, no heavy draperies. This makes a "live" acoustic environment in which reflected sound reinforces the output of your stereo system. As a result you can cut your power needs back by about 50 percent. But if your environment is acoustically "dead," with wall-to-wall carpeting, heavy draperies, overstuffed chairs and sofas, pillows and wall hangings, it will soak up sound and place an extra burden on your amplifier. In that case tack an extra 50 percent onto your power budget. Open doorways, by the way, should be regarded as an equivalent area of sound-absorbent curtain.

Now let's consider your musical tastes. So far, our calculations are based on the assumption that you play symphonic music at reasonably loud levels. But if Wagnerian thunder or massive organ sound is your special delight, you can make those crashing tonal cascades even more thrilling if you up the power figure so far determined by about 30 percent. On the other hand, if your predilections run exclusively to string quartets or small jazz combos, you can cut 30 percent off the calculated wattage. But why limit yourself so? Ideally, a good sound system should be able to reproduce any kind of music convincingly, whatever its original setting.

If you find that your musical inclinations and domestic surroundings decidedly make you a high-powered type, ask yourself one more question: Can your speakers handle all that power pumped into them by the amplifier? Make sure that the power-handling capacity of your speakers (as stated in the manufacturer's specifications) equals or surpasses the per-channel rating of your amplifier. Otherwise, those hoped-for fortissimi may drive your speakers into distortion or blow-out.

Juggling all these figures in your mind can be confusing. To simplify things, consult the power chart on pages 89–90, which summarizes these numerical matters.

POWER AND DISTORTION

So far in this discussion we have been speaking of power ratings in watts as if this were a clearly defined measurement. Unfortunately it is not. Power in itself means very little in terms of performance. What counts is *undistorted power*. As an amplifier approaches its power limit, distortion also rises with output. Consequently, an amplifier's power rating must always be specified relative to its distortion at the stated power level. As yet, there is no general agreed standard of measurement that specifies the distortion level at which the wattage is measured. Consequently, power-rating figures have been juggled by some advertisers in a way that is virtually fraudulent. The Federal Trade Commission is currently considering standards for power measurement in the audio industry, but until these standards are in force the wary buyer must acquaint himself with the various ways in which manufacturers mask rather than reveal the true capabilities of their products.

This is not to imply that all manufacturers of sound equipment are crooks. On the contrary, the more reputable component manufacturers have formed an association, the Institute of High Fidelity (IHF), largely to set honest standards for performance measurements. Most component manufacturers adhere to these standards and list the letters IHF with their specifications to signify that they do.

POWER BUDGET PLANNING CHART
for average-sized rooms

LISTENING ROOM	LOUDSPEAKERS	RECOMMENDED WATTS PER CHANNEL
"Live" Acoustics		
Little sound absorption.		
Tile or linoleum floor, no	High-efficiency	8
rugs, small curtain area,	Medium-efficiency	12
smooth walls.	Low-efficiency	15
Average Acoustics		
Curtains and some carpets,	High-efficiency	10
some but not much uphol-	Medium-efficiency	15
stered furniture.	Low-efficiency	20

LISTENING ROOM	LOUDSPEAKERS	RECOMMENDED WATTS PER CHANNEL
"Dead" Acoustics Extremely high sound absorption. Wall-to-wall carpet, heavy draperies and curtains, stuffed chairs, couches, pillows, wall hangings.	High-efficiency	15
	Medium-efficiency	25
	Low-efficiency	50

For rooms greater than 30 feet in length or width, increase the recommended wattage by 20-50 percent.

Fudging the Specs

Even the IHF rules can be considered slightly equivocal, for they don't specify at what level of distortion power measurements must be made. For example, one manufacturer's specification might read "20 watts (IHF) per channel at 1 percent distortion," while another maker might rate an identical amplifier at 2 percent distortion, so as to be able to list a higher wattage figure. Hence you cannot directly compare IHF wattage ratings unless they are specified for the same percentage of distortion. Most manufacturers, however, rate their amplifiers at 1 percent or less. If they don't, the specifications usually tell you so.

But among the large firms making ordinary radios and phonographs, fidelity is hardly the chief concern. Setting up their own rules through the Electronic Industries Association (EIA), these companies rate amplifiers at a whopping 5 percent distortion. As a result, the power claims are impressive, but the sound is not. What's worse, the 5 percent distortion allowed by the EIA is never explicitly stated. "It would confuse the customer," says an EIA spokesman, "and since all our measurements are made at 5 percent, it would be redundant to say so." It is somewhat difficult to follow such corporate logic. After all, it is less confusing to state the facts than to conceal them. Also, it can't be redundant to provide information not given elsewhere, and one hopes that the federal government will share this view. Meanwhile, the best rule to follow is to buy only sound equipment that has specifications stated according to the more meaningful IHF standards, as long as a distortion figure is supplied along with the wattage rating.

The matter of power rating is made still more ambiguous by two different measurement standards employed for quality components: "continuous power" or "rms power" and "music power." The actual difference between these two ratings can be as much as 30 percent of the rated power. Continuous or rms power is the more rigorous test, yielding the more conservative rating. It measures the power output of the amplifier over sustained periods of time. Music power, by contrast, measures the power output the amplifier can reach during short bursts of loud sound. To understand why these two are different, it is necessary to go into some of the engineering factors that determine an amplifier's power capacity.

Power Supplies

Two design factors determine the wattage an amplifier can deliver. One is the circuitry of its output stage; the other is the so-called power supply. As its name implies, the power supply of your amplifier or receiver is responsible for supplying the proper voltages and currents (power) to the various electronic circuits. Most of the resources of the power supply are drawn by the amplifier's output stages, which have the task of raising the audio signal to a strength sufficient to drive loudspeakers.

It is not always easy for the power supply to keep up with the demands of the rest of the amplifier, for high-powered amplifiers do not draw a constant amount of current from their power supplies. Their demand varies according to the level of the signal being amplified. The louder the sound the speakers are called upon to make, the more current is required. Deep organ notes, crashing kettledrums, heavy piano chords—all these represent high concentrations of power, and they are the real tests of an amplifier's power supply. For example, a kettledrum beat or a loud orchestral chord calls for many times the current drawn during quieter passages. No matter how well it performs during quiet passages, the amplifier cannot cope with these musical emergencies unless its power supply has plenty of reserve current instantly available at such moments. Moreover, the power supply must be able to recover quickly from these sudden drains on its reserves so that its voltage level is restored before the next loudness peak comes along, which may be only a fraction of a second later. All this comes under the heading of power-supply regulation, a factor particularly important in solid-state equipment.

Power-Supply Regulation

A power supply for a transistor amplifier consists mainly of three elements: (1) a power transformer to change 120-volt AC house current to the various lower voltages required within the amplifier; (2) a rectifier (usually a solid-state bridge) to convert low-voltage AC to the DC needed by the amplifier; (3) filter capacitors to smooth out remaining traces of AC ripple and also to act as a storehouse of extra energy for moments of high power demand.

Transformers and capacitors capable of providing ample current reserves for high-wattage output stages tend to be both bulky and expensive. That is why in the interest of compactness and economy some manufacturers are forced to skimp a little in this department. Naturally, this narrows the safety margin by which the amplifier surmounts critical moments. Depending on your taste in music, how loud you play it, and the efficiency of your speakers, there may or may not be audible deterioration of sound. A poorly regulated power supply may provide enough current for a rating of, say, 40 watts per channel with a momentary loud note at mid-range frequencies, but when a long, loud bass passage comes along, the amplifier performs no better than a unit with a 15- or 20-watt rating.

This discrepancy raises the question: Should this amplifier be called a 20-watt amplifier or a 40-watt amplifier? In other words, should it be rated on its continuous power or its short-term power? As usual, the logic of the promotion department prevailed in this matter. The short-term standard was adopted and decorated with such fancy names as "music power" or "dynamic power." Though the purpose of such subterfuges is to make a piece of equipment seem better than it is, the music power rating is fairly useful. At least you can compare the relative ratings of two amplifiers or receivers, as long as the power measurements are made at the same distortion level.

While you cannot directly compare music power ratings and continuous power ratings, the latter is by far the more rigorous and unequivocal test. Wherever you see a component specification listing "continuous power" or "rms power" you can just about take it for granted that it is an honestly designed piece of machinery and that the manufacturer feels he has nothing to hide. In some cases of high-quality equipment you will find that the spec sheet lists both music power and continuous power.

Other Obfuscations

It also makes a difference whether power measurements are made with *all channels operating,* or are made with only one channel in operation at a time. Since the demands made on the power supply by one channel will diminish the resources available to the other channel, only the all-channel figure indicates the amplifier's true performance capability.

The maximum power available from transistor amplifiers also depends upon the impedance of the speaker load attached to them. For example, a certain amplifier may develop 90 watts per channel into an 8-ohm speaker, 120 watts into 4 ohms, and only 50 watts with a 16-ohm speaker. Note how the manufacturer rates such an amplifier. That's a clue about his philosophy. Most manufacturers would list this as a 90-watt amplifier since most speakers have an impedance of 8 ohms. Very aggressive salesmen would quote the 4-ohm figure, calling the amplifier a 120-watter, even though it could not develop this much power with most loudspeakers.

Another trick is to add up the wattage of both channels for the power rating. An amplifier might be honestly described as "30 watts per channel." But by adding up the channels, the same unit could be called a 60-watt amplifier. It is expected that the FTC will put a stop to this. The FTC may also eliminate such quasi-fraudulent designations as "peak power" or "peak music power." These are nothing but a brazen and arbitrary doubling of the music power rating. Such figures convey nothing but a grossly inflated notion of an amplifier's merit.

In sum then, the present situation concerning amplifier power ratings reflects not so much technical facts but a condition endemic in our consumer society: the baleful corruptions compounded of manufacturers' greed and consumers' gullibility. The foregoing explanations are intended to help you see through the disguises and obfuscations so as to recognize the real merits of a product. The cheerful fact is that, despite the prevalence of misleading advertising, truly fine equipment is available in all price and power brackets.

6
Amplifiers II—Distortion and Separation

Aside from the basic matter of power, the performance of an amplifier depends on such factors as distortion and separation, which should be plainly stated on the specification sheet. We have already indicated that power ratings are meaningful only in relation to the amount of distortion generated at a given power level. After all, the amplified signal must correspond precisely to the waveform of the original input, or else the sound will be falsified. Low distortion is therefore an indispensable requirement for good sound equipment.

Only within recent years has distortion been acknowledged as the prominent touchstone of quality in sound reproduction. Earlier in the electronic quest for musical realism—roughly during the adventurous decade following World War II—any system pretending to wide-range frequency response was considered "high fidelity." As long as the bottom bass (anything below

100 Hz in those days) and that tinkly treble (perhaps 10,000 Hz) came through, listeners who had never heard reproduced sound like this before were amazed, impressed, and happy. Little did it matter to them that those "deep" lows were boomy and/or muddy and that the metallic harshness and screech of the highs ruined all chance of tonal realism.

Since then we have become more sophisticated in our demands. Wide-range frequency response is now pretty much taken for granted in quality equipment, and the emphasis has shifted to keeping the tonal timbre as natural as possible—in short, to minimizing distortion. Complete elimination of distortion still eludes even the most ingenious of electronic wizards; in fact, the characteristics inherent in any circuit or device make absolute perfection theoretically impossible. But the long fight against musical falseness has taken us astoundingly close to the limits of what is possible. Where obvious distortion once made violins sound as if they were made of metal rather than wood and turned trumpets into raucous kazoos, today's best components reduce distortion to virtually unnoticeable levels. Whatever distortion is audible on modern top-grade equipment very likely originates in the program source (records, tape, or FM transmitter) rather than in the components themselves.

Unfortunately, not all audio equipment on the market meets these exacting standards. Although the performance of few of them is marred by blatant distortion, their tonal impurities are usually quite subtle and for that reason all the harder to recognize. At first, you may not notice any distortion at all. But after an hour or so of attentive listening, you may find yourself getting fidgety, vaguely uncomfortable, less responsive to the music, and increasingly disposed to "turn the damn thing down—or off." What happens is that marginal distortion, too slight to register consciously, builds up subliminally to create a psychological effect known as "listener fatigue." By contrast, top-grade equipment lets you listen for many hours without waning pleasure. One might think of this curious reaction to flawed sound as the mind's instinctive protest against the sonic adulteration of music.

Fortunately, detecting distortion need not be left to the subconscious. Accurate physical testing methods have been worked out to assess distortion.

TYPES OF DISTORTION

Distortion can be defined broadly as anything that keeps the music that reaches you through your loudspeakers from being a flawless replica of the live music you might have heard in the presence of the performing artists. But in the technical sense distortion usually means the kind of tonal falsification stemming from changes in the waveforms of electronically reproduced sounds. Three principal types of such distortion—harmonic, intermodulation, and transient—plague audio designers.

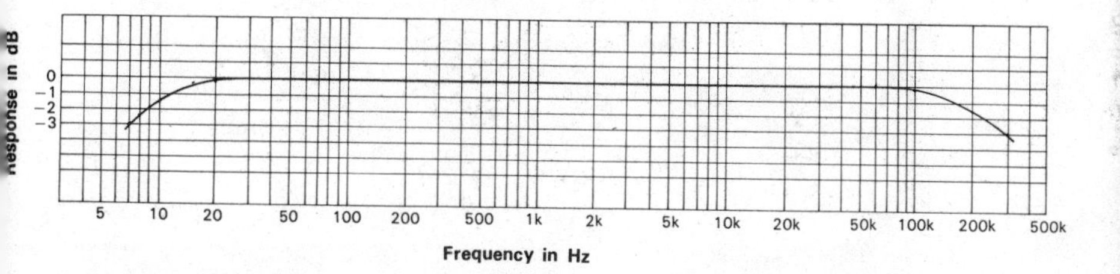

Amplifier frequency response graph as it might appear in the specifications of a component. In this excellent unit, response is essentially flat throughout the audible range, falling off only below and above the audible limits. (The abbreviation k stands for 1000. Thus 1k = 1000 Hz.)

Harmonic Distortion

The first, harmonic distortion, takes its name from the harmonics, or overtones, that are produced by an instrument in addition to its basic pitch. It is these natural overtones, or harmonics, that give an instrument its distinctive tone color. Suppose you are playing a 1000-Hz note on a trumpet. What emerges from the instrument is not a pure 1000-Hz tone, called the fundamental, but a mixture of the fundamental tone with the harmonics or multiples of that frequency, such as 2000, 3000, and 4000 Hz. What determines the character of the trumpet sound is the exact proportion in which these overtones appear—that is, their strength in relation to each other and to the basic pitch. These harmonics are not distortion; they spring from the basic sonic nature of the instrument.

Harmonic distortion enters the picture if your sound equipment distorts the waveform and adds overtones of its own, thus altering the proportions of the original overtone pattern. These spurious overtones, generated within the equipment, are then superimposed on the music and thus change the essential natural coloration of the various instruments. No amplifier, cartridge, tuner, or speaker is entirely free of this unwelcome tendency to add its own impromptu frequencies to the music. But the object of good audio design is to minimize such additions.

The usual way to measure harmonic distortion in the laboratory is to feed into the amplifier a test tone consisting of a single pure frequency. A filter connected at the amplifier's output then suppresses this particular frequency. Whatever other frequencies still remain in the amplifier output must be the products of harmonic distortion. The strength of the harmonic-distortion signal is then measured and expressed as a percentage of the amplifier's output at a given power level. Thus, a typical specification may read: "Total harmonic distortion (THD) .3 percent at rated output." Distortion readings should also be stated at 1-watt output to indicate the amplifier's performance at low volume.

Harmonic distortion is the least obtrusive of the three types, yet for all its subtlety it may well foil the attentive listener's attempt to tell a Baldwin from a Steinway or to compare the characteristic tonal aura of the Vienna Philharmonic with that of the Philadelphia Orchestra. For such fine-eared listening to be possible, THD above 200 Hz should not exceed .5 percent. Sensitivity to distortion varies with the frequency and the har-

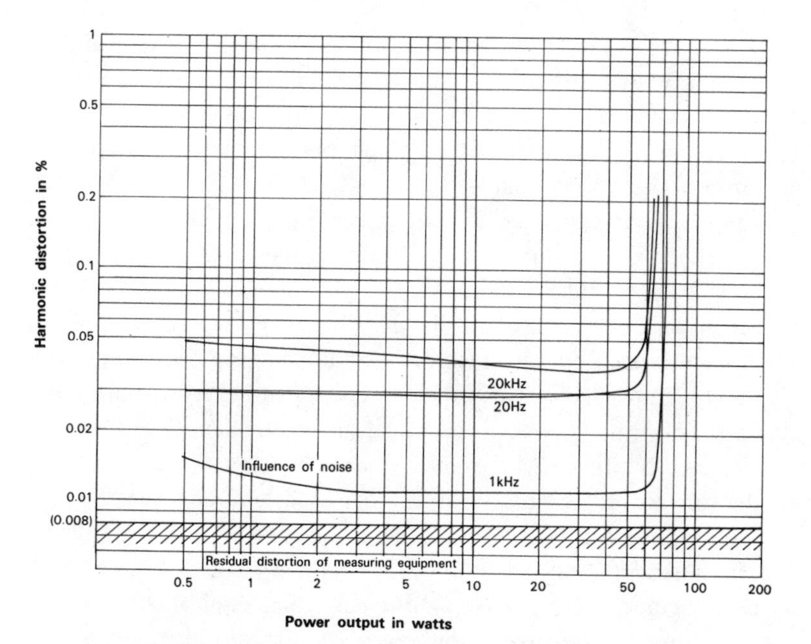

A distortion graph, plotting harmonic distortion (in percent) against power output in watts. As the power limit of the amplifier is reached at about 50 watts, distortion can be seen to rise sharply on both channels. The test tones are 20 kHz (= 20,000 Hz) and 1 kHz.

monics (second, fifth, tenth, and so forth) involved. Fortunately for sensitive listeners, audio designers have succeeded in reducing harmonic distortion in high-quality amplifiers to inaudible levels.

Intermodulation Distortion

The most disturbing of these three distortion types is unquestionably intermodulation distortion (IM). The raucous harshness that it adds to music is sadly common in ordinary consoles and record players, although, to a certain extent, the limited high-frequency response of cheap equipment prevents the worst part of the distortion from being audible. In component equipment, IM distortion is the predominant cause of listener fatigue.

Like harmonic distortion, IM is caused by nonlinearity, which is to say that the shape of a waveform coming out of an amplifier is not the same as that which went in. IM distortion occurs in nonlinear amplifiers when two or more tones pass through at the same time—as is usually the case in music. What happens is that the various frequencies interact with each other (intermodulate) and thereby produce illegitimate offspring. Suppose a 60-Hz note and an 8000-Hz note are traveling together inside the amplifier. By the time they reach the output they will have produced at least two additional notes the composer never wrote. One will be equal to the sum of the original two (8060 Hz), and the other will equal their difference (7940 Hz). The fact that these IM-engendered tones have no harmonic relationship to the original tones is what especially enhances their sonic irritation quotient. What is more, the two superfluous tones also interact. And when there is a whole orchestra fiddling, blowing, and banging away, and the electrical equivalents of the sounds are furiously and spuriously interacting, the result is a musical mishmash.

To keep such disorderly conduct under surveillance, audio engineers perform IM tests. The test consists of putting two tones through the amplifier and then taking a census of their unwanted by-products. Both test tones are pure when they enter the amplifier. At the amplifier output, frequency filters suppress the two parent tones; what remains is distortion, which is measured and expressed as a percentage of the total output. Under standard test conditions, the two test tones are 60 and 6000 Hz, applied at an intensity ratio of 4 to 1.

In a good amplifier, the IM rating is usually kept below 1 percent at full-power output and at all other signal levels. After all, it is just as important for the music to be clear and true in soft passages as in loud ones. That is why many manufacturers now also specify IM distortion at the 1-watt output level in addition to full-rated output. A curve showing the percentage of

IM distortion at all power levels is the best way to evaluate an amplifier's performance in this respect.

With IM remaining below the 1 percent limit at all power levels, listening fatigue is not likely to be a problem, even after several hours of continuous and attentive listening, such as hearing a complete opera performance. If the other elements in the sound-reproduction chain—recording, pickup, and speaker—are also reasonably free of intermodulation distortion, the music is reproduced with an aesthetically rewarding aura of naturalness and clarity.

Transient Distortion

Having dwelled on the subject of distortion, I now propose to end the topic with a bang. In acoustic parlance, a "bang" is a transient, something that comes and goes quickly. Transients are sounds bursting forth suddenly at high loudness levels and breaking off just as fast. Music is full of transients, such as the crash of the stick on the drumhead, the clangorous moment of contact between a pair of cymbals, or the hard impact of piano hammers against the strings where the score says fortississimo. Yet, in the sum total of orchestral sound, these dramatic collisions, however spectacular, are less pervasive than their subtler relatives, the tongue against the reeds of woodwinds. These, too, are transient sounds, lasting in their critical phase only a small fraction of a second. In the orchestral aggregate, these unobtrusive transients contribute vitally to the texture of sound.

Transients are difficult for amplifiers to reproduce. They fall victim to transient distortion, a deplorable process that can turn a sharp, hard click or snap into a fuzzy "thunk." The reason is that electronic circuits, like speaker cones, are afflicted with inertia. They have trouble starting or stopping swiftly enough to reproduce the transient's wave shape accurately. When the sudden bang comes along, it causes a fast-rising electric pulse. Engineers, describing the visual pattern presented by such a pulse on the oscilloscope, say that it has a "steep wavefront" or a "fast rise time." This means that the amplifier must jump to the signal's peak correspondingly with virtually no inertial delay. And when the bang is gone, the amplifier must go back into neutral as promptly as it left.

Only very good components can do this. Others tend to round off the steep wavefronts or persist in jiggling away in spurious oscillation long after the transient signal has ceased. The ear perceives such transient distortion as a sonic pall cast over the whole texture of the sound, clouding the musical detail. The sound gets soggy, and its washed-out quality drains the music of its tension and excitement.

Unfortunately, there is no altogether satisfactory way of measuring transient distortion. A very few manufacturers specify the "rise time" of their amplifiers, that is, the amount of time it takes to reach a specified output level. Measured in microseconds (millionths of a second), this is a pretty good indication of the amplifier's reaction speed. Like all test tones, these are simplifications of the more complex waveforms generated by musical instruments. Consequently, these tests provide only presumptive evidence of an amplifier's transient characteristics. In my opinion, the human ear, listening for sharpness of tonal definition and clarity of texture, still remains the most reliable test instrument for transient distortion.

S/N AND SEPARATION

Of the remaining amplifier specifications, we must consider signal-to-noise ratio and separation. Signal-to-noise ratio, often abbreviated S/N, is sometimes listed on the spec sheet as "hum and noise." This ratio expresses relative amount of interference with the signal. In the language of electronics, noise is any kind of unwanted sound that intrudes into, or interferes with, the desired signal. Perhaps the most consistently unappreciated pleasure of high fidelity is that all these forms of noise are held to a minimum by good equipment and that the music emerges from a silent background. The signal-to-noise ratio is expressed as the loudness difference (in decibels) between the desired signal (usually measured at the amplifier's full-rated output under test) and the interfering noise. The specification "hum and noise—60 db" means that hum and other noises are at a level 60 db lower than the musical signal reproduced at full output power. A rating of 60 db is good—the higher the figure, the lower the noise.

Separation refers to the amount of right-channel signal spuriously sneaking over into the left channel, and vice versa. (In quadraphonic systems separation also refers to keeping front and rear signals apart.) Separation is tested by feeding a test-tone signal into one channel of a component and measuring how much undesired signal appears at the output of each other channel. The difference between the two channels is expressed in decibels—the greater the number, the better the separation. You can check the separation yourself with a test record that contains test tones in one channel at a time. To the question "How much separation is necessary for good stereo performance?" engineers tend to reply, "The more the better." The majority of today's components are easily capable of all the separation that program material seems to require.

Adequate separation is very important in the mid-range and upper mid-range frequencies, say 400 to 8000 Hz, not only for spatial localization of the instruments, but for reproducing the acoustic environment of the original performance. On the other hand, a component's separation at frequencies below 400 Hz is not particularly significant since the natural laws of acoustics prevent accurate localization of sound in that range anyway. At frequencies of 10,000 Hz and above, it becomes increasingly difficult to maintain separation in a hi-fi component because of the capacitive-coupling effect that causes interchannel feed-through of high-frequency signals. However, since the amount of energy (and program material) at these frequencies is quite small, the deterioration of separation in that range is not very noticeable.

It doesn't seem to be generally understood that although a component may be capable of separation in excess of 20 or 30 db, the program material itself may not have that much channel-to-channel isolation in it. In fact, it is safe to say, at least with classical material, that the better the recording the less obvious and startling the separation will be. The greater-separation-than-life recordings made during the early days of stereo have given way to more natural-sounding, smoothly blended sound. The object of modern recording is to spread the sound evenly between the speakers. An analogy can be drawn between listeners who like exaggerated, unreal separation and those who prefer heavy, larger-than-life bass response. Their aim, apparently, is to improve the original performance, not duplicate it. They are certainly entitled to this approach, but they should not be under the impression that it has anything to do with high-fidelity reproduction.

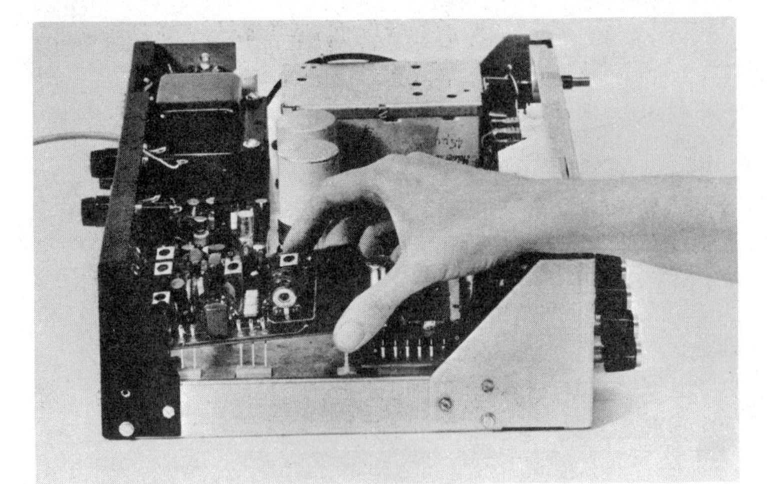

The solid-state, fully transistorized innards of modern amplifiers make it possible to pack ample power into compact space. Plug-in, modular construction minimizes service problems. (*Photo: H. H. Scott, Inc.*)

7
Knobs, Buttons, and Switches— Your Personal Sound Control

With its profusion of knobs, buttons, slides, and switches, the front panel of an amplifier or receiver evokes intimidating visions of a jetliner cockpit. The comforting fact is that in day-to-day listening only a few of the many controls are ever used. But the others are there for good reasons. They give you a vast range of expanded options far beyond the possibilities of ordinary phonographs and consoles. And once you acquaint yourself with the purpose and the logic of their function, you can derive a great deal of pleasure from their intelligent use.

To start with, any amplifier has a *selector switch*. As its name implies, it lets you pick the program source you want to hear: radio, record player, or tape. Each position is clearly marked, and the function is self-evident. The function of the *volume control*—sometimes labeled "level" or "loudness"—is equally obvious. Usually, however, you will also find a switch called

Front panel of AM/FM stereo receiver shows multiplicity of controls. Push-button controls at right simplify operation. (*Photo: H. H. Scott, Inc.*)

"loudness compensation." This requires some explanation, but its proper use can contribute greatly to your listening enjoyment.

LOUDNESS COMPENSATION

In the interest of natural sound, it is best to hear music at the same relative volume as in the concert hall. This doesn't mean that your speakers must pump out as much sound as a whole orchestra. After all, your living room is a lot smaller than the concert hall, so you need less power to attain the same sound intensity. What matters is that the level of sound at your ears should be about the same as it would be if you were sitting in a good seat at a concert. If you can play music at this volume, your tone controls should be in the "flat" (straight up) position.

But sometimes irate neighbors pound on the walls to make you turn down the volume. Something odd happens then. The music doesn't just get softer. Its tonal balance also changes. The low notes seem to drop out, leaving the sound texture thin and tinny, and the sparkle of extreme highs suddenly seems dulled.

The orchestra suddenly sounds as if cellos and basses had gone on strike, leaving the sound texture minus its due weight and sonority.

If the music itself is soft, this subjective loss of lows is part of the intended musical effect. The psychological quality of a pianissimo partly depends on the way the ear registers soft sounds. But if the original music is loud, the same subjective factor distorts its psychological effect if the volume is cut down in playback.

But why, one might ask, does a change in volume also produce a change in quality? Why doesn't the music sound the same only less loud? The answer is that the frequency response of the ear itself differs at various volume levels. At moderately high sound levels in the range of 60 to 80 db (such as the sound of an orchestra playing fairly loud) the frequency response of the ear is almost flat; that is, all frequencies—high, middle, and low—are heard with a relative loudness proportional to their actual physical energy. But if the level is reduced, the higher frequencies, and especially the lower ones, no longer sound proportionally as strong as those in the mid-range (about 1000 to 5000 Hz). People noticing this tend to believe that their sound equipment is at fault in that it doesn't put out enough treble and bass at low volume. Rarely do they suspect that the phenomenon is caused by subjective factors in their own hearing.

The causes for this nonlinear response of the ear are still a mystery, but the effects have been carefully studied in audiometric measurements made by two American scientists, Harvey Fletcher and W. A. Munson, some thirty years ago. Their data, expressed in the so-called Fletcher-Munson curves, form the basis for the design of the loudness-compensation circuits.

The principle is simple: At low volume settings, the loudness control boosts the bass and the treble by approximately the same amount that the human ear suppresses it. At higher volume settings, it has no effect at all. The net result, ideally, is that the subjective impression of the music remains realistic, regardless of the volume at which you play it. You switch on the loudness control and, presumably, you can achieve both natural frequency balance and peace with the neighbors. The hitch is that not everyone's hearing conforms perfectly to the Fletcher-Munson average, and the loudness-compensation circuit is designed for this "average" hearing characteristic. This is akin to fitting everyone with an average-size shoe.

If, for these reasons, the results don't seem satisfactory, you can modify the loudness compensation with the regular treble and bass controls, to the point where the music sounds full and properly balanced to you. If your amplifier lacks loudness compensation, you can use treble and bass controls to get the same effect. Just add bass boost and a very slight amount of treble boost until the music sounds full and natural.

Some unregenerate audiophiles contend that the illusion of orchestral fullness attained through loudness compensation is not realistic, and it must be admitted that loudness compensa-

Extreme simplicity of controls characterizes this compact amplifier, which is also available in kit form. (*Photo: Dyna Co.*)

tion certainly is no substitute for playing music at its natural volume. But at least it permits you to hear all the notes without having to shake the walls down.

BALANCE ADJUSTMENT

In addition to volume and loudness controls, you also need some way to adjust the loudness of each channel separately, so that the sound from left and right speakers arrives at your listening post in proper balance. This is done by means of the balance control. As you turn the control toward the right, the right speaker gets louder. Turn it to the left, and the left speaker gets louder.

Some listeners wrongly believe that simply leaving the balance control centered (pointing straight up) assures correct balance. This is true if all the speakers are equally efficient at all frequencies and if signals of equal strength are fed to them. In the early days of stereo, many people also thought that an area in front of and equidistant from both speakers was the only possible listening location, and they would huddle on an imaginary center line between the two speakers. The fact is that the balance control, properly used, permits you to adjust the balance for many other locations in your room. If you sit closer to the left speaker, it will naturally seem louder than the right. You can compensate for this on most amplifiers by a slight twist of the balance control toward the right. If you sit closer to the right speaker, you turn the control toward the left. The idea is to adjust the control so that both speakers sound equally loud from where you sit.

Try the following: Set your amplifier to mono. This assures that all speakers will get the same signal. Then sit in your favorite chair, close your eyes, and ask a member of your family to turn the balance control slowly back and forth. At a certain control setting, the music will seem to emerge from an area between the speakers. This indicates the optimum balance setting for your particular listening spot. By following this procedure you also automatically take into account other variables that may affect stereo balance, such as uneven gain in the two amplifier channels (a common but usually unsuspected failing) and differences in the acoustic efficiency of the two speakers.

A few amplifiers have a special device for balance setting that reverses the phase of one stereo channel so that the sound is partially canceled by phase interference when the two channels are in balance. The clearly audible "null effect" then pinpoints the position of the control at which electrical stereo balance is attained. But this achieves acoustical balance only if the two speakers are identical. One can also adjust stereo balance with the aid of test records containing special signals for this purpose. But the simple procedure outlined above can be carried out without any equipment other than your own two ears, which are what you are trying to satisfy in the first place.

TREBLE AND BASS CONTROLS

Many listeners are knob-shy. They are somehow afraid to fiddle with the controls and to experiment with the effects they can achieve by their use. This applies especially to the *tone controls.* The term "tone control" recalls the pre hi-fi era, when only a single knob was provided to cut back on the amount of the higher frequencies, or treble tones, that were permitted to reach the listener's ear. On modern high-fidelity instruments, the tone-control action is considerably more sophisticated. At least two separate controls are provided, one for treble, another for bass. This enables the listener to select the specific tonal range he wants either to emphasize (boost) or to attenuate (cut) in order to obtain the most satisfying musical balance.

The bass control usually acts on frequencies from about 250 Hz downward, i.e., in the range from the middle register of the cello or trombone all the way down to the lowest audible notes.

The treble control usually acts on frequencies reaching from the upper range of the piccolo and fiddle to the highest overtones that define the timbre of the various instruments. The middle frequencies—roughly the notes around high C (approximately 1000 Hz)—remain unaffected by either control.

Though the exact degree and frequencies at which treble and bass controls increase or decrease sound in their respective ranges differ among various makes and models of equipment, nearly all of them operate similarly as far as the user is concerned. Turning the bass control clockwise makes the low notes more prominent in relation to the others; turning the control counter-clockwise weakens the bass. The treble control works the same way in the upper range. When the controls are in neutral position—usually pointing straight up in a twelve o'clock position—they have no effect whatever on the tonal balance. Engineers call this the flat position because, when represented as a frequency graph, it results in a flat line showing neither boost nor attenuation at any frequency. In some expensive equipment, there is even a switch to remove the tone controls from the circuit entirely. This lets you switch instantly from flat position to any favorite tone-control setting.

The widespread timidity about the use of tone controls may have its origin in the misleading notion that treble and bass controls should always remain in a neutral or flat position, pointing straight up. With the controls so set, bass and treble are neither boosted nor depressed, and the amplifier yields the flat response beloved of theoreticians. With bass and treble unaltered by the controls, highs and lows emerge from a well-designed amplifier in the same relative proportions as they went in. It is as if the controls weren't there at all. This is dandy, and may well assure maximum musical realism—if the following four conditions are met: (1) the listening-room acoustics treat all frequencies alike; (2) the loudspeakers also have a flat frequency response; (3) the engineering of the record, tape, or broadcast being reproduced has preserved the natural musical balances between highs and lows; and (4) the music is heard at concert volume.

Under such ideal conditions, tone controls are indeed superfluous, and we would be absolutely right in leaving them flat. But our world is full of acoustically idiosyncratic living rooms, weak-bottomed and shrill-topped loudspeakers, hoked-up records, inept broadcasting engineering, and irascible neighbors with an explicit aversion to concert volume. Thus tone controls, far from being needless knobby appendages on your sound equipment, can provide the means of coping with less-than-ideal circumstances.

The acoustics of your living room are quite likely to give undue emphasis to some parts of the total frequency range, while at the same time de-emphasizing others. Smallish rooms, for example, tend to make it difficult to reproduce the low bass frequencies. Hard plaster walls and picture windows can overstress the treble frequencies, making the music bright to the point of shrillness. Conversely, heavy draperies, rugs, and upholstery soak up the highs and can make the music dull and lifeless. But it takes only a slight twist of the tone controls to counteract most of these shortcomings.

Or suppose you have a pair of small speakers whose response falters at low frequencies. Without those deep bass fundamentals that support the whole structure of orchestral sound, the music seems insubstantial, lacking that fullness and weight essential to the very concept of symphonic scoring. With tone control, however, you need not tolerate this deficiency. A slight touch of bass boost—say, turning the bass control from the twelve o'clock to the two o'clock position—may supply at least part of those missing lows, restoring to the orchestral sound some of its natural depth. The better the speakers, the more kindly they take to frequency corrections by tone controls.

Conversely, whatever harshness is produced in the high register by imperfect speakers can usually be tamed by nudging the treble control slightly in the counter-clockwise direction—just enough to take the edge off the sound, but not the bloom. Related shortcomings of discs and broadcasts can be similarly overcome.

Changing the Sonic Image

Treble and bass controls are not just for correcting the listening room's acoustic deficiencies and making up for less-than-perfect speaker response; they are also very useful for correcting sonic shortcomings of recordings and broadcasts. A case in point: I have in my disc library a recording of a Buxtehude cantata, a magnificent performance. But much of its musical meaning is lost because the bass line, the vital foundation of this fine musical structure, almost disappears when the disc is played with the tone controls in their normal flat position. To make matters worse, the soprano screeches. It isn't her fault, but rather the engineers', who, in a misguided effort to make the record sound brilliant, also put an edge on her voice. Why the bass is weak is anyone's guess. It could be that the bass fiddle was too far from the mike, or perhaps the engineers deliberately weakened the bass to make it easier for inexpensive phonographs to track the grooves. The problem with this disc is how to let the musical beauty shine through despite the technical blemishes. The tonal cosmetics needed are just what tone controls are meant for. A slight diminution of treble takes the rasp off that angelic soprano, and a fairly hefty dose of bass boost (turning the knob to the three o'clock position on my preamplifier) restores the music's aesthetic balance by giving the lows their necessary weight.

Fortunately, shortcomings of this sort are getting rarer these days as the average quality of recordings improves. But live broadcasts of symphony concerts and opera performances are still beset at times by similar imbalances of highs and lows. There are many possible causes for this: A concert hall acoustically ideal for listening is not necessarily ideal for broadcast purposes; architectural quirks may prevent microphones from being placed to best sonic advantage; and the circumstances surrounding a live broadcast do not always provide time or space for experimentation to find the best possible microphone locations. Again, the tone controls are the listener's ready recourse for bringing highs and lows—and thereby the whole orchestral and vocal texture—closer to a natural, realistic balance.

To some degree, these controls also let you "argue" with the conductor. For instance, where the conductor subdues the lower strings to obtain a lighter sound, you can countermand him with your bass control and change the orchestral coloration toward a heavier, darker hue. Or, by accenting the treble, you can bring into clearer focus certain details in the score that the conductor subordinated to the overall orchestral blend. Whatever its effect on your ego, such artistic free enterprise has questionable musical merit if used excessively. Exaggerated and willful tonal changes will certainly falsify the music. In a way, it is like looking at a Renoir with sunglasses on. Musical meddlers who habitually crank up the bass all the way and boost the treble "to add brilliance" will get music that sounds like a stomach growl against a counterpoint of shattering glass. Tone controls are best used with discretion and a light touch.

Most bass and treble controls operate over a fairly broad frequency range, with their effectiveness concentrated near the high and low ends of the audio spectrum. On some amplifiers more elaborate versions are now available that divide the frequency spectrum into five or seven regions, each of which can be separately controlled. These devices let the listener tailor the frequency response of his system to complement the frequency response of his room. He may apply a judicious amount of suppression at those spots in the spectrum where his room tends to create a bass boom or a harshness in the highs. Or he may boost those frequencies that his room tends to swallow up.

These devices grew out of sound reinforcement methods employed in auditoriums and churches, and though they have been simplified, many of them can still be considered as professional apparatus, somewhat complicated to operate. For one thing, it is quite possible to do more harm than good and only an ear trained in critical perception can be trusted as a guide in setting the controls.

FILTER CONTROLS

For the less technically oriented, at least two functions of these variable filter systems can be performed more easily and inexpensively on most amplifiers by controls known as *scratch and rumble filters.*

Most surface scratch and distortion caused by worn and dirty records is above 7000 Hz in frequency—a range that contains many of the overtones but none of the fundamental frequencies of music. The scratch filter eliminates the noise simply by lopping off that range. A well-designed scratch filter provides a sharp cut-off in response above a certain frequency, rather than a gradual dropping off. It is intended to clip off unwanted noise while leaving most of the music intact. A sharp-cut filter usually requires a fair amount of additional circuitry, involving feedback and extra amplifying stages. This extra circuitry is the reason why high-quality scratch filters are seldom found on inexpensive amplifiers and receivers.

Of course, hiss and other high-frequency noise reduction can also be achieved by turning down the treble control. However, this produces a gradually declining roll-off rather than an abrupt cut-off. As a result, more of the music is sliced off along with the noise. Even the best scratch filter curtails to some extent the range of overtones and thereby degrades the brilliance and naturalness of sound. The scratch filter must therefore be viewed as a necessary evil, needed to salvage musical enjoyment from old or mistreated records that would otherwise be unlistenable.

The *rumble filter* provides a corresponding cut-off function at the low end of the audio spectrum to eliminate unwanted sounds at the bottom of the scale. Turntable rumble is a frequent complaint of this sort, for poor turntables are notoriously prone to vibrations that, when picked up by the cartridge and amplified, sound vaguely like indigestion. But one should not always blame the turntable for such sonic indiscretions. They are sometimes caused by acoustic feedback, a condition afflicting improperly installed turntables. The loudspeakers' acoustic output feeds back to the turntable through air, floor, and cabinet. In some cases, the rumble filter reduces the noises by cutting off response below 50 Hz. Yet here, too, the music suffers a slight amputation, losing the added richness that is the hallmark of extended bass response. With bass-shy loudspeakers, this loss may not be noticed. But with speakers capable of reproducing low bass below 50 Hz, the effect of a rumble filter is quite apparent.

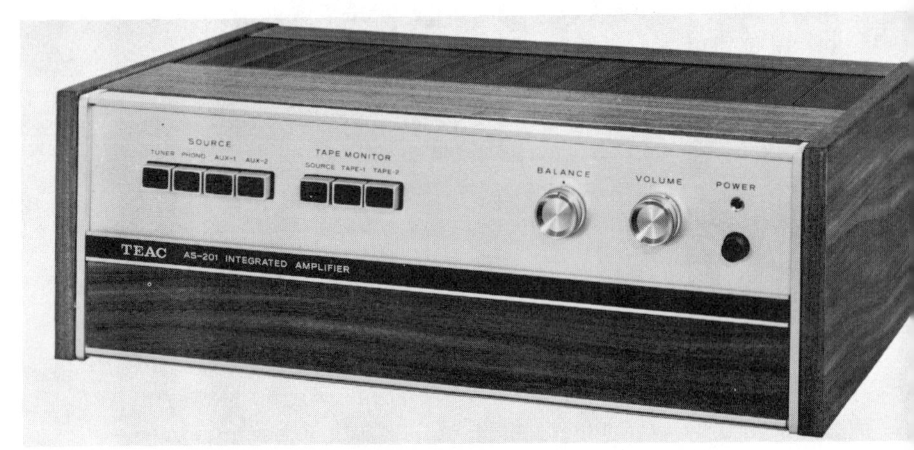

To avoid confusion among the many controls, this amplifier puts only the most frequently used controls out in the open. All others are concealed beneath a hinged panel below. (*Photo: TEAC, Inc.*)

Both scratch and rumble filters are essentially cosmetic devices designed to mask imperfections and shortcomings in records or turntables. They should be used only as an emergency measure if extraneous noise is too distracting. It is best to eliminate part of this noise before it appears—by keeping your records clean. In addition, make sure that your turntable is operating properly and that it is installed with the specific shock mounts recommended and supplied by its manufacturer.

All of these variable controls, volume, balance, and tone, may appear on your amplifier front panel either as conventional knobs or as "sliders." Though the latter do have definite advantages in professional recording studio equipment, they neither contribute to nor detract from the value of home equipment.

MODE, MONITOR, AND SPEAKER SWITCHES

Virtually every stereo amplifier has a *mode switch.* Its function is to set up the amplifier for stereo or monophonic operation. One prevalent attitude toward this control is to ignore it, leaving it set on stereo even when playing mono records. This does no harm if the mono disc is in good condition (with quiet surfaces) and the turntable rumble is low. But if the mono record is full of distracting ticks and pops that seem to bounce between the channels, or if it is old and has groove dimensions not particularly suited to today's smaller styli, performance can be noticeably improved by flipping the mode switch to mono. The reason for this improvement is that by switching to mono the effect of vertical stylus movement is canceled out. On mono records all the music is embodied in the side-to-side wiggles. Up-and-down stylus movement represents nothing but noise. Setting the switch to mono therefore cuts the noise without hurting the music. Sometimes the mono setting also results in noise reduction on weak tuner signals by producing phase cancellation of the noise present in the two channels. While you lose the stereo effect, switching to mono may be the only way to get a listenable signal at all from a weak station.

On some amplifiers, the mono position is designated as A+B (or L+R), signifying the blending of channel A (left) and channel B (right). Such amplifiers usually also have special mode-switch positions for channel A or channel B alone. In these positions the input to either one of the channels is applied to both speakers. This provides a handy way of checking a stereo-signal source. For example, by switching alternately from channel A to B you may be able to pinpoint noise or distortion as being mostly in one of the two channels. Another possible use of these switch positions is as auxiliary program selectors if the regular selector switch cannot accommodate all program sources hooked to the amplifier. For example, in addition to a stereo tape deck and a stereo FM tuner hooked to the regular inputs provided for them, you might connect an additional AM tuner to the auxiliary input on channel A and perhaps a mono cassette machine or some other mono input device to channel B. With the input-selector switch on AUXILIARY, the mode switch can then be alternated between the channel A and channel B positions to select one of the two additional program sources. In a similar fashion, the mode switch may serve to select either one of the two tape tracks from a stereo tape recorder that does not have provision for the playback of only one track at a time. Some mode controls also have intermediate degrees for partially blending the two stereo channels. This can be useful if your loudspeakers are spaced very far apart, or when you play stereo records with exaggerated separation. But for all normal purposes, this control is not necessary.

Another control frequently found is a *speaker muting control.* This lets you switch off the speakers when you're listening through earphones, an important feature when you want to enjoy your music without disturbing others. Many amplifiers also provide the option for hooking up an extra pair of speakers to provide music in another room. In that case, the amplifier also features a *speaker selector switch* to let you pipe the music into either or both rooms.

If you are planning to add a tape recorder or cassette machine to your system, a control called the *tape monitor switch*—found on nearly all the better amplifiers—will prove extremely useful. The tape monitor works like a railroad switch: it selects which of two paths the signal will follow. In the normal position (usually marked either "source" or "normal"), it acts as if it weren't there at all. The signal follows its normal path through your input to the speakers. In the other position (usually marked "tape"), the monitor switch ignores the signal coming from the selector and instead feeds the signal from the tape recorder to the speakers.

The purpose of all this is to permit you to make instant comparisons between the source signal (radio or record) and the tape recording you make of it. By flicking the monitor switch back and forth, you can compare the source you are recording with the actual tape you're making of it. This way, you can make sure your recording is all right. And if it isn't you can immediately make the necessary adjustments. (Of course, this requires a high-quality tape recorder with separate heads for recording and playback.)

REAR-PANEL CONTROLS

Some audio fans, beguiled by the array of knobs and switches on the front panel of their amplifiers or receivers, pay scant attention to the controls located on the rear. Occasionally these adjustments are ignored altogether, with the result that the user gets less than the best possible performance from his equipment. Rear-panel controls, in general, are intended to be set only once and then left in that position. Their relative inaccessibility, therefore, is no disadvantage. In fact, some of these controls are not operated by switches or knobs but have slotted shafts that must be turned with a screwdriver. This prevents their being accidentally dislocated once they are adjusted.

The most common rear-panel control is a switch or knob for setting the sensitivity of the phono input: this matches the input characteristics of the phono preamplifier to the strength of the signal put out by the cartridge. Cartridges, depending on make and model, differ in their outputs. A high-output magnetic cartridge could overload the preamplifier input stage of your transistorized receiver or amplifier, which would result in distortion during loud passages. But if the phono-sensitivity switch is put into the "high" position (i.e., the position intended for a high-output cartridge), a resistance is added to the signal path to prevent possible overloading and distortion.

The literature supplied by the manufacturers of your equipment should give the information you need to make this adjustment properly. If this is for some reason unavailable, set the switch so that the position of the volume control is high for your preferred listening level, but not high enough to introduce preamplifier noise (hiss). Only if you are not getting enough volume even at a fairly advanced setting of the volume control should you set the rear-panel phono-sensitivity switch to "low." (Note: "high" and "low" in this context refer to the output of the cartridge, not the sensitivity of the input stage.)

Individual input-level controls are sometimes provided for each pair of input jacks, or for all but the phono inputs. You adjust these so that all your signals will be heard at the same level, to avoid the annoyance of abrupt volume changes when you switch from radio to records or to tapes.

The rear panel is also, of course, where most of the connections are: high-level inputs, low-level inputs, and outputs. The outputs include the speaker terminals, the signal take-off for reel tape recorders or cassette machines. Other inputs to the amplifier may be marked: "Tape," "Tuner," "TV," or simply "Aux" (for "auxiliary"). You can use them interchangeably. Not so the phono inputs, which are designed for the specific purpose of connecting the record player. On all quality amplifiers, this phono input is designed to accept the output from a "magnetic" cartridge (see Chapter 9). Some very cheap amplifiers do not have inputs for magnetic phono cartridges. They are designed for "ceramic" cartridges. Since this type of cartridge, used widely in ordinary nonhi-fi equipment, does not meet stringent audio standards, an amplifier that does not accept magnetic cartridges is automatically ruled out from consideration as a quality instrument.

A few recent amplifier models sport such unusual extras as microphone inputs alone with a separate volume control for the microphone that lets you blend music (say, from a record) with your own voice. This is handy for making tapes of your own sing-along or for recording soundtracks for your home movies or slide shows.

Most amplifiers also feature a simple AC power outlet at the rear. There you can plug in other components (turntable or tape machine) without having to snake the power cord all the way to the nearest wall outlet. Some of these AC outlets are on all the time, as long as the amplifier is plugged in. Others switch on and off with the amplifier power switch.

As we have pointed out before, most amplifiers today are integral one-piece units in which control functions and amplifying functions are combined in a single component. Many of them are also combined with a radio tuner so as to make a "receiver." The most expensive and most powerful amplifiers, however, come in two separate pieces. One contains the controls and is called the preamplifier; the other is called the power amplifier and contains the output stages. It is because in very powerful amplifiers the output stages are necessarily big and bulky that they are housed separately from the control functions. But the control functions, as such, remain the same as in other designs and the foregoing discussion applies to all kinds of amplifiers.

COMPONENT KITS

Many good amplifiers and receivers are available in kit form. This raises the question whether it might be wise for you to enter the higher strata of fidelity by way of a kit project. Obviously, there are advantages to this approach. Short of outright larceny, I know no more economical way to acquire an amplifier than to build it from a kit. Since labor is the most expensive single ingredient in an amplifier (or tuner) you can save some 30 to 40 percent of the price of comparable factory-finished equipment by doing your own assembly work. Such savings have enabled many high-fidelity fans to acquire equipment of a quality that otherwise would have been beyond their means, and many kit builders have found that, apart from saving money, they really enjoy putting their own equipment together.

In recent years, kit building has become far simpler than it used to be. In the early days of high fidelity you had to be reasonably familiar with the basic anatomy of components if you wanted to build them yourself, but most kits sold today are so designed that even those who have never held a soldering iron can easily learn to put them together. In many modern kits all possible doubt about what goes where has been removed, and some include even such details as connecting cables that are color-coded and precut to the right length. And if you manage to make an error despite all this, a self-checking routine that is part of your instructions helps you spot and correct your mistake. Virtually nothing is left to chance or your own judgment.

Of course, previous experience is helpful, but it is no longer indispensable. The experienced kit builder may complete the job faster, but if the novice takes his time and works carefully, he can end with a component that works just as well. This is not to say that kit building is easy for everyone. If experience is no longer a requisite, other qualifications are—notably patience, a certain degree of handiness, and the ability to do a job systematically. Possibly the greatest help of all is a temperament that will allow you to sit still for several hours and work methodically, step by step. Those kit builders who try to outsmart the instructions and invent shortcuts of their own usually wind up with some interesting-looking wire sculpture. But it won't play music.

A soldering iron, a pair of pliers, and a screwdriver are all the equipment you really need; and, depending on the kind of component you are building, the job may take anywhere from eight to forty hours. A power amplifier, which has relatively few parts, can be completed in two evenings, and some of the newer integrated amplifier kits have lately been simplified to the point where they take only a little more time. Receivers are considerably more complex and may take more than forty hours to put together. I would not recommend such a project to any but experienced kit builders.

8
Four-Channel Systems

The pros and cons and maybes of quadraphonic sound have already been discussed in the opening chapter. For the present it merely remains to indicate the different types of four-channel equipment now available. If, like many audiophiles, you decided to stick with standard stereo, read no further. Just skip ahead to Chapter 9.

The first and obvious question is: How do you get four channels of program material in the first place? In other words, what four-channel sound sources exist today? Basically there are two sources: four-channel tape and four-channel records.

FOUR-CHANNEL TAPE

By far the simplest way to make a four-channel recording is on open-reel tape. Standard reel tape already has four tracks, two in each direction. Lately a number of companies have been making tape recorders on which all four tracks can be recorded and played simultaneously. Fidelity is fine, but these machines are quite expensive, prices starting in the vicinity of $600. Prerecorded four-channel tapes to play on these machines are avail-

A professional-quality tape deck for recording four channels simultaneously. Note the four separate recording-level meters. Optionally, four-track recorders of this type also can function in standard stereo, two tracks at a time. (*Photo: Sony/Superscope*)

able, but prices are high and the repertoire is very limited. Four-channel open-reel tape thus can only be considered a rather esoteric venture.

A cheaper four-channel tape system has gained some popularity. Taking advantage of the eight-track pattern in the tape cartridges commonly used for automotive stereo, a number of companies now offer so-called Quad-8 or Q8 cartridges in which four tracks are recorded and played simultaneously. Many companies now offer equipment for playing or recording these Q8 cartridges, but none of these units meets stringent standards of high fidelity. There are inherent limitations in the present design of eight-track cartridge equipment that limit its usefulness for serious music listeners. Besides, the available repertoire in this format is mostly pop stuff.

FOUR-CHANNEL RECORDS

The best chance for widespread acceptance of four-channel sound is with four-channel records. The problem here is how to get the two extra channels into the record groove. Audio engineers have come up with two workable answers. One is a technique known as matrixing by which front and back channels on each side are blended into one. That way, the four channels are squeezed down into two on the disc. Then, in playback, a matrixing adapter or decoder attached to the amplifier (or built into it) reverses the process. It separates the blended channels again, feeding front and rear signals on each side to their respective speakers. Columbia Records and a number of other labels are now offering an expanding repertoire on such "SQ" records. When these records are broadcast over FM stereo stations, the decoder acts in the same way, so you can get four-channel music off the air if the station you are listening to is broadcasting such material. If you don't have a decoder and four-channel equipment, these broadcasts will come through like standard stereo.

A four-channel amplifier with separate sound-level meters for each of the four channels. (*Photo: Heath Corp.*)

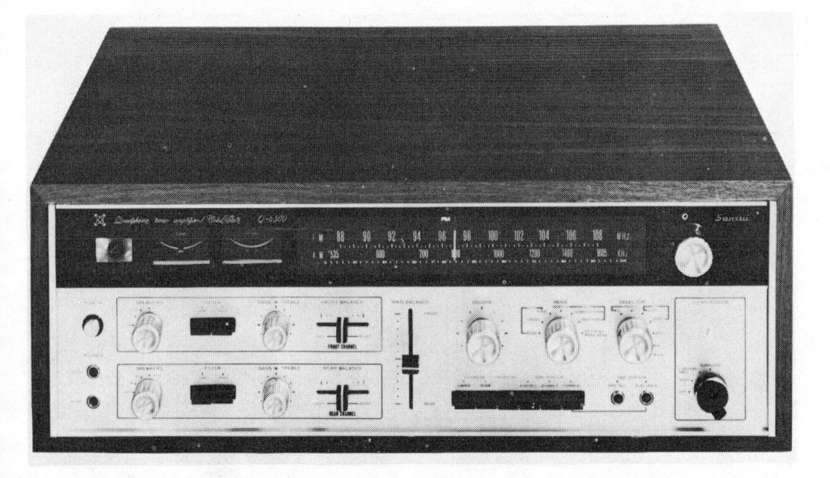

A four-channel stereo receiver, combining a radio tuner with four-channel amplification and built-in decoder. (*Photo: Sansui Corp.*)

You can also play these "SQ" records on any standard stereo system for ordinary two-channel listening.

RCA is plugging a rival system which by some fancy engineering legerdemain actually squeezes four separate channels into a single record groove. A special phono cartridge is needed to play these records. The advantage of these discs is that they provide somewhat better channel separation than the matrixed "SQ" system. The drawback is that they might wear out faster and that playing time per side is somewhat limited.

CONVERTING STEREO TO QUAD

Any standard stereo system can be converted to four-channel sound by adding a decoder, two extra speakers, and an extra amplifier for the two rear channels. Many models are now available that combine the decoder and the extra amplifier in a single unit.

A four-channel converter. Adding this unit to a standard stereo system provides four-channel capability. (*Photo: Sansui Corp.*)

FOUR-CHANNEL CONTROLS

Most four-channel adapters have master volume controls that regulate the volume of all four channels simultaneously. In addition, they feature a control for adjusting the relative volume of the back channels relative to the front channels. They also feature a side-to-side balance control for the rear channels. The function of these controls is analogous to the corresponding controls for standard stereo discussed in the preceding chapter.

9
Record Players—Music Preserved

Around the turn of the century, Ambrose Bierce compiled his pet notions in alphabetical order. His list, published under the title *The Devil's Dictionary,* defines the phonograph, then in its infancy, as "a machine to revive dead noises." Bierce, who shortly afterwards vanished in the Mexican jungle, would have relished the irony of the fact that the phonograph grew up to be the exact opposite of his definition: an instrument to assure that some "noises" shall not die.

The phonograph has made possible the preservation of music not merely in vague paper notation but in the sound itself. This is plainly a matter of life and death, for music on paper is preserved by being mummified in symbols. By contrast, music on records is preserved, as it were, while still breathing. Music used to be instantly perishable. Inevitably it would vanish with its own sound. Today, thanks to recordings, it spans distance as well as time. No matter where, no matter when recorded music was played, the recording always puts it in the present, always on the spot. Today, as most music is heard at a distance in time and space from the actual performance, recordings form the bridge between music and the listener.

The most common kind of recording is on discs. More re-

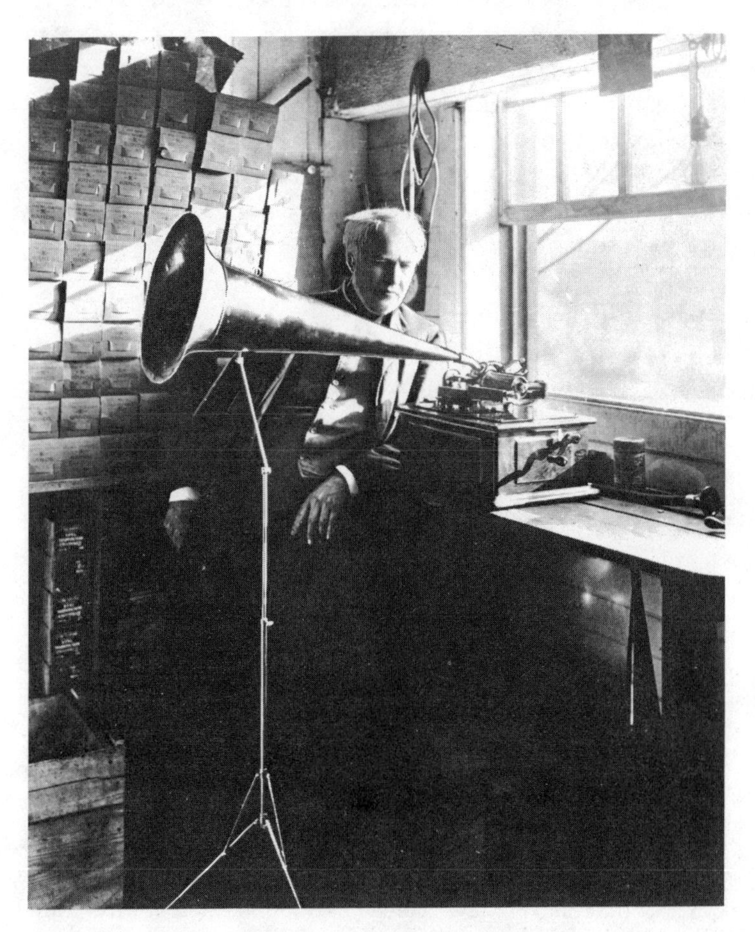

Thomas A. Edison, inventor of the phonograph, with one of his early models.

cently, magnetic recording on tape has also become quite popular. We shall speak of them in a later chapter. For the present, we are concerned mainly with phonograph records and the best way to play them.

THE BASIC TYPES OF RECORD-PLAYING EQUIPMENT

There are three basic types of record-playing equipment: manual turntables, automatic record changers, and automatic turntables. All three types have the same basic parts: a motor-driven platter that rotates the record at the proper speed, a cartridge whose stylus (or "needle") traces the wiggly contour of the grooves and converts the motion into an electric signal, and a tone arm to hold that cartridge in place above the groove.

Manual Turntables

By itself, the word "turntable" can mean simply a motor-and-platter system to which you can affix a separate tone arm. More commonly, "turntable" refers to a turntable and tone arm combination. Either way, the term usually describes a manual system in which you pick up the arm and set it down at the beginning of the groove to play a record.

Record Changers

A record changer adds two other elements: a spindle to hold a stack of records and drop one at a time, and a mechanism that sets the arm down at the beginning of each record and lifts it again at the record's end. Changers provide hours of uninterrupted music without any attention on your part, and then turn themselves off when the last disc has played.

Despite these extras, most changers cost *less* than manual turntables. Built for the mass market at the lowest possible prices, these inexpensive changers seldom elicit all of the sound from the record grooves and often wear out records after just a few plays.

This record player is equipped with a manual turntable with precision tone arm.

A fully automatic turntable with provision for changing records

A turntable with automatic arm-positioning and shutoff

Automatic Turntables

The third alternative—the automatic turntable—is simply a changer built to high-fidelity standards (and costing proportionally more). Automatic turntables are usually supplied with both a tall spindle for automatic record changing and an alternate short spindle for greater convenience when playing just one record at a time.

CHOOSING YOUR TYPE OF TURNTABLE

For the hard-bitten purist, the manual turntable with a precision tone arm is probably still the preferred choice. But I must confess that my own preference is for a good automatic turntable, simply because today's precision machines position the tone arm more gently on the record than I ever managed to do with my rather fumbly fingers. And an automatic turntable with a good tone arm causes no more record wear than a manual model. Indeed, the automatic turntable is the best way to protect your records and the very vulnerable phono cartridge from the casual clumsiness of inexpert users. They never need to touch the tone arm at all.

In other aspects of performance, today's best automatic turntables—such as those made by Dual, Miracord, or Norelco, and the top models of the Garrard line—are equal to most professional-type manual turntables. And in the price range from about $100 to $130, automatic turntables represent very good value in terms of the cost-performance ratio. Excellent manual turntable-tone arm combinations are available for less, with some outstanding models selling for about $80.

TURNTABLE AND TONE ARM PROBLEMS

Turntables and tone arms differ from all other audio components in one respect. Every other component handles some form of signal or sound, but the turntable and tone arm are supposedly silent partners in the enterprise of sound reproduction. Yet, like many such partners, they may wield a not-so-silent influence over the whole operation, vitally affecting the musical outcome. An inferior turntable ad libs an unscored tremolo or adds so much rumble that the music sounds as if accompanied by a distant thunderstorm.

Turntables on ordinary phonographs are usually beset by these weaknesses to a distressing degree. To keep quality turntables free of them takes expert design, precision manufacture, and rigid inspection. This explains why good turntables, despite their essential simplicity, tend to be relatively expensive.

Turntable Rumble

Vibration presents the thorniest problem. All rotating machinery shows a bit of roughness in the running, and the trick with a record player is to keep it from becoming audible. This is quite difficult because the whole record-playing system acts as a kind of vibration detector. As long as the vibrations originate in the record groove, you've got music. But if they come from the turntable itself, you've got rumble. And although designers can't get rid of rumble entirely, they try to keep the ratio between music and rumble overwhelmingly in favor of the music.

This ratio, expressed in decibels (db), is the so-called rumble rating—the most important of turntable specifications. The specs may tell you, for instance, that a given turntable has a rumble rating of −35 db, which means that its rumble is 35 db less loud than a standard recorded test tone. This figure, if obtained in accordance with the standard test procedures of the National Association of Broadcasters, represents the minimum requirement for broadcast-station turntables as established by the NAB, and it indicates that the turntable is very silent indeed. But if the figure is not specified as being made according to the NAB standard, it may be meaningless. Some of the best home turntables have still higher "minus figures" and therefore even lower rumble ratings.

Most rumble frequencies lie between 30 and 50 Hz, coinciding with the lowest reaches of the orchestra. The exact frequency depends upon the source of the rumble—the turntable bearing, the drive system, or the motor. If your loudspeakers are capable of reproducing low bass, you may find your music accompanied by an unscored growling obbligato competing with the bass line and beclouding the upper frequencies. Compounding the problem is the fact that rumble frequencies, like all extreme lows, may overtax your amplifier and speakers, driving them into distortion.

Flipping on your amplifier's rumble filter has the doubtful merit of therapy by amputation. True, it eliminates the rumble, but it also deprives you of the full low bass that is one of the touchstones of high fidelity. So, if you have a persistent rumble, getting a better turntable is the only really effective countermeasure.

Banishing Acoustic Feedback

Before rushing out in search of a suitably silent turntable, you'd better make sure that those distracting growls you hear are really rumble. Another culprit, acoustic feedback, could be to blame. This occurs, as we mentioned earlier, when vibrations from the speakers get back to the turntable and pickup, to form a dog-chasing-his-tail loop. The ensuing sound, being low-pitched, is easily mistaken for rumble. Such feedback often stems from setting the speakers and the turntable on the same table or shelf. It is common in consoles, where turntable and speakers are often mounted in close proximity. Even if the turntable and speakers are separated at some distance, the speaker's vibrations can get back to the turntable through the floor.

For a quick test for acoustic feedback, set the tone arm on a record while the turntable is not running. Turn up the bass controls a bit beyond their normal position and then slowly turn up the volume control while tapping the turntable base. If a growling roar gradually builds up in your speakers, you've got a case of acoustic feedback.

A thick foam-rubber mat under your record player and speakers may help. Or you can move your turntable to another area farther from the speakers. If rumble still persists when the record turns, it's a safe bet that you need to repair or replace your turntable. Most turntables are supported by soft springs that help soak up vibrations transmitted through furniture or floor. This is why turntables should be mounted with the hardware recommended by the manufacturer, even when they are built into custom cabinets.

On some turntables the motor, turntable, and tone arm are supported by a separate sub-frame, suspended on springs beneath the upper surface of the turntable cabinet. Such systems usually have good built-in resistance to acoustic feedback. The Norelco automatic turntable and the AR manual turntable are outstanding examples of this type.

Speed Constancy

Another vital requirement for turntables is speed constancy. If the turntable motor doesn't drive the platter evenly, the result is a chugging motion that causes the music to waver. This is known as flutter, a quivering tremolo especially noticeable on long-held notes of fixed-pitch instruments, such as piano and organ. Slower speed variations of the turntable produce a pitch wobble descriptively called "wow."

In quality turntables these defects are avoided through the use of (1) highly specialized motors that maintain their speed and torque over large variations of line voltage and (2) drive systems that efficiently couple the torque of the motor to the rotating platter. The driving force may be applied through a belt, a soft drive wheel mounted directly on the motor shaft, or through an intermediary "puck" that picks up the motor drive and transmits it to the turntable platter.

All of these systems are capable of excellent rumble and wow and flutter specifications. Specifications for wow and flutter are given in percentage figures, which express the turntable's maximum fluctuation from the desired standard speed (33⅓, 45, or 78 rpm). On a good turntable, for example, you may find wow and flutter as low as .1 percent, which produces no audible pitch wobbles.

Some turntables permit slight variations on the standard record speeds by means of so-called pitch controls. These controls let you speed up or slow down a record by about 3 percent of the standard speed and thus change the pitch and tempo of the recorded music within narrow limits. This is handy if you want to play a musical instrument along with the record, for that way you can "tune" your record player to play exactly in pitch with your piano, trumpet, or clarinet. To get the pitch control set accurately for standard speed, some turntables have a built-in stroboscopic speed indicator. If they don't, you can buy a strobe disc as an inexpensive accessory to calibrate your turntable for the right speed.

MOTOR TYPES

Advertisements for turntables often specify the kind of motor used to drive the platter. A few words will suffice to distinguish the different types.

Induction motors, used on less expensive turntables, are the least desirable, because their speed changes if your power line voltage fluctuates. Synchronous motors, which lock onto the frequency of your electric house current, are more trustworthy. The expensive "hysteresis synchronous" motor used to be highly praised for speed constancy, but it has been matched in performance by the better examples of today's ordinary synchronous motors.

The servo motor is a new and promising development, maintaining its speed accuracy through a type of feedback. The motor itself runs on DC, and its speed is proportional to the voltage supplied by the control circuit. The control circuit, in turn, senses the actual speed of the turntable, instantaneously raising the voltage if the speed is too low, lowering the voltage if the speed is too high. In addition to greater potential speed accuracy, the servo system has two other advantages: it can operate on either 60-Hz current in the United States or 50-Hz current abroad without requiring parts replacement. Also, its motor turns at a lower speed, which makes rumble frequencies lower, putting them below the range of hearing.

TONE ARMS

As for tone arms, they seem deceptively simple at first glance—just a stick on a swivel, holding the cartridge and supporting it on its way across the disc. But today's stereo cartridges are pretty demanding about the kind of support they get while their styli travel through the record grooves. The stylus assembly of a high-compliance cartridge is so flexible that a clumsy, heavy tone arm would simply squash it; even the best cartridge can't live up to its specified promises if the arm won't "play along." The tone arm, then, is an active working partner of the cartridge. How well it acquits itself in this partnership depends on four main factors: stability, tracking accuracy, resonant properties, and friction.

Unlike the other three specifications, a tone arm's stability cannot be expressed by a simple number, yet it represents a vital requirement: the ability of the arm to keep a correct and constant downward force on the stylus at every moment of playing. On low-fidelity phonographs the arm is usually balanced by a spring whose effect is to pull the arm upward to counteract the weight of the arm and cartridge pressing downward on the record. However, the pull of the spring makes the arm jumpy. Heavy footsteps, passing traffic, or a warp on the record may cause the stylus to skip grooves. The stylus, the record, and the listener's nerves are the chief victims of such poor tone-arm design.

Quality tone arms are never balanced by an upward-pulling spring. Instead, they are balanced by some type of adjustable counterweight at the rear. Springs, if they are used, serve only to provide the necessary downward tracking force. In many designs a rear counterweight is set to balance the front part of the arm so that the stylus pressure is zero. By a calibrated adjustment of a spring or counterweight, the exact amount of downward force required is then added for optimum performance from the cartridge. Balance of this kind lends such stability to the arm that the tracking force remains constant regardless of floor vibrations or slight accidental jolts. Result: the arm won't jump. Moreover, the centers of gravity of the forward and rear sections of a quality arm are so positioned that the stylus tends to remain centered in the groove even when the turntable is tilted.

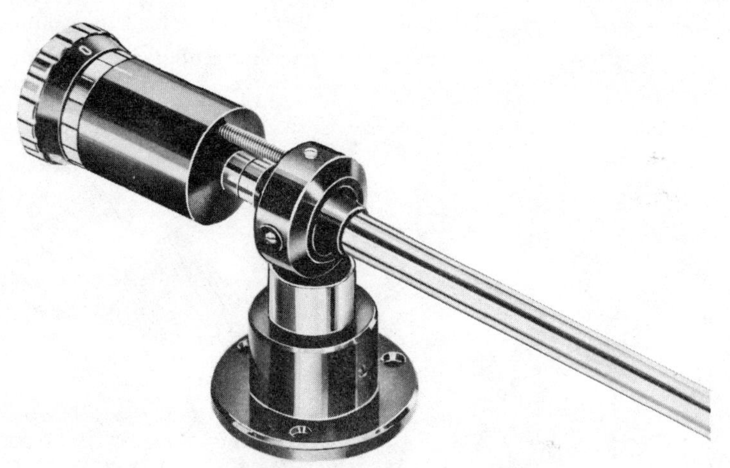

Precise balancing and low-friction bearing are essential for a good tone arm. This detailed view of the tone arm mounting shows the precision-machined parts. (*Photo: Ortofon, Ltd.*)

Anti-Skating Devices

Several new tone arms include additional balancing mechanisms called "anti-skating" devices. The purpose of these devices is to assure that the stylus pushes against both sides of the groove wall with equal force. Because of a slight amount of frictional drag on the stylus as it travels through the groove, the stylus tends to lean a little harder against the inner groove wall than the outer one. Anti-skating devices overcome this imbalance.

Tracking Error

The path traversed by the tone arm across a record is as cunningly calculated as the orbit of a spacecraft. Ideally, the tone arm should travel across the record in a straight line from outer edge to record center to keep the cartridge correctly aligned with the record grooves (tangent to the grooves) all the way from beginning to end. But the arm can't travel in a straight line because it swivels on a pivot. As a result of its curved path, the cartridge changes its angle with respect to the grooves as it scans the face of the disc. The number of degrees by which the actual cartridge angle differs from true tangency is called the tracking error.

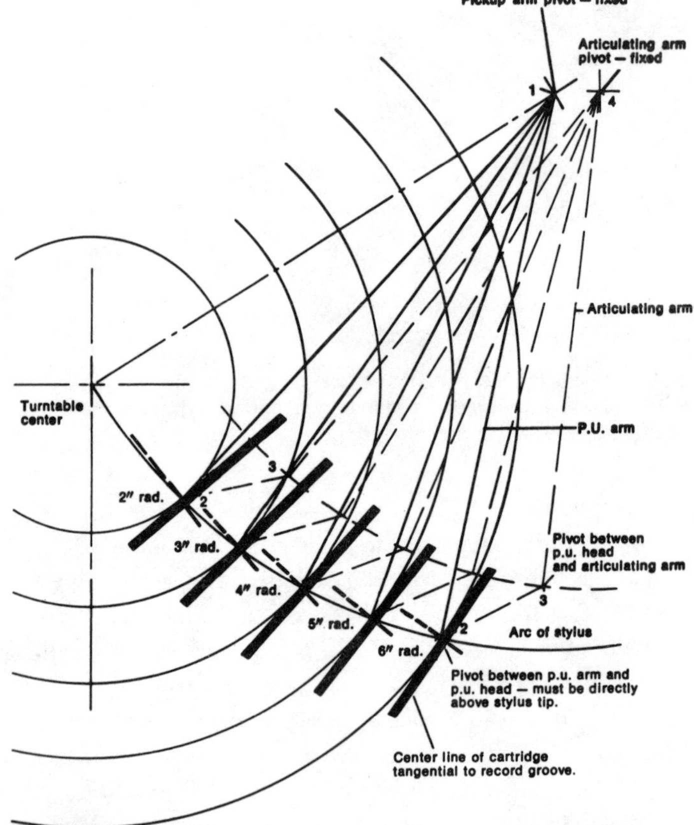

Concept drawing illustrating maintenance of accurate tracking angle. Tangency of tone arm should be maintained regardless of diameter as the tone arm moves from the rim toward the center of the disc. (*Garrard, Ltd.*)

One way of reducing tracking error is to lengthen the arm. Mathematically, if the arm were infinitely long, the tracking error would be zero. But tone-arm designers have found ways to reduce tracking error even when working with arms of less than infinite length.

One might expect that true tangency could occur at only one place on the record. But by angling the head of the tone arm toward the center of the record and by mounting the arm so the cartridge travels toward a point slightly off the record center, it is possible to achieve not one, but two true-tangent points along the arc of tone-arm travel. Having two points of zero tracking error brings down the average error to below 2 degrees at all points on the disc. Some designers feel that it is more important to design tone arms for minimum overall playback distortion than for minimum overall tracking error. This is to say that they are willing to tolerate a slightly higher tracking error in the outer grooves, where tracking distortion is less audible, if, by doing so, they can reduce the tracking error at the inner diameters, where tracking distortion is more noticeable.

All this is not merely drawing-board sophistry, for accurate tracking does provide an audible advantage. If the cartridge is mounted askew rather than tangent to the groove, the stylus cannot respond equally to the wiggles on both sides of the groove. The result is distortion that is especially noticeable on the inner grooves where the musical waveforms are more tightly packed together.

A few tone-arm designers have therefore gone to great lengths to reduce tracking error virtually to zero. Two approaches are commonly used. One is actually to move the cartridge in a straight line across the disc. The other is constantly to vary the angle between the tone arm and the cartridge, so that the cartridge remains tangent to the groove no matter where the arm is pointing. Both approaches have technical difficulties to overcome, but both can be made to work well, though not inexpensively.

Resonance and Friction

Aside from tracking and balance, two additional factors—resonance and friction—greatly affect the quality of a tone arm. Audio designers take great pains to control tone-arm resonance so that the arm will not itself vibrate with the frantic dance of the stylus in the record groove. If the arm resonates at any of the musical frequencies, its own vibrations are piled on top of the musical signal. The result is a tonal hash of intermodulating frequencies, along with added record wear as arm vibrations beat the stylus against the groove. To forestall this, the natural resonance of a well-designed arm should be considerably below the lowest notes likely to be encountered in the music, preferably below 15 Hz. Resonance control is also the reason why some arms are made of wood (an inherently well-damped material), why you may find viscous damping in the pivots of some arms, and why some tone arms also have a semi-elastic linkage between the arm and its counterweight.

The need for low friction stems from the fact that any mechanical resistance to the motion of the arm pulls on the stylus and distorts the signal, especially in stereo. Besides, the more force needed to overcome tone-arm friction, the more stylus pressure it takes to keep the stylus centered in the groove. To permit modern high-compliance cartridges to track at pressures of 2 grams or less, the arm must be virtually frictionless. To accomplish this, tone-arm designers employ precision bearings for both the lateral and vertical pivots of the arm. While the bearing friction is seldom stated numerically in the specifications, it can be judged by the feel of the arm as you move it about. It should move smoothly and evenly with a minimum of resistance.

In a sense, the turntable and the tone arm are merely a kind of stage setting for the real performer in record reproduction—the phono cartridge.

CARTRIDGES

The cartridge—the little device at the tip of the tone arm—is where the music really starts. It is the gateway by which the music enters the sound system. The cartridge is the vital first link in the chain of sound, and as the originator of the signal, it occupies a strategic position. If it doesn't read out the sound correctly from the record groove, the music gets messed up at its source. No matter how good your amplifier or your speakers are they cannot correct the faults introduced at the start by an inferior cartridge.

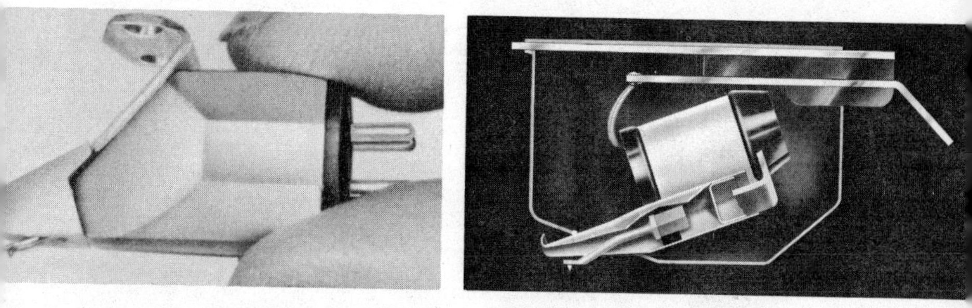

Left: Although it is the smallest of all components, the phono cartridge (or pickup) exerts a major influence in the quality of record reproduction. (*Photo: Empire Co.*)

Right: Greatly enlarged view of cartridge in plastic case reveals working element and shows the delicate stylus assembly with its guard shell and diamond tip. (*Photo: Benjamin-Elac*)

Next to the speakers, the cartridge has the most notable effect on the kind of sound you hear. And like your speakers, the cartridge adds its own distinctive coloration to the sound. Speakers and cartridges are alike in this respect for the same reason. They both are so-called transducers.

Transducers are middlemen between separate provinces of nature—the mechanical and the electrical. Where the speaker transduces electrical energy into mechanical motion, the cartridge changes mechanical motion into electrical energy, translating the varying waveforms physically molded into the record groove into corresponding variations of electrical voltage. The accuracy of this translation largely determines the fidelity of the reproduced sound. To be accurate, the stylus must follow with utmost precision the complex waveforms on the disc. In stereo, it must wiggle up and down as well as sideways in a dipping and swaying journey through the groove. To reproduce a 15,000-Hz overtone, for example, the stylus must travel through 30,000 hairpin turns per second. And to follow the violent undulations representing loud bass notes, it must endure accelerations greater than those experienced by astronauts during blast-off. Yet, throughout this wild ride, the stylus must never lose contact with the groove, not even for a microsecond. Otherwise, the result would be an ear-grating harshness of sound on loud or high-frequency passages.

Compliance and Mass

The tracking behavior of a cartridge can be inferred from two commonly given data: compliance and dynamic stylus mass. Compliance measures the force necessary to push the stylus from its neutral (center) position—that is, the force the record groove must exert against the stylus to make it follow the musical waveforms. The higher the compliance, the less force is needed. Because high-compliance cartridges yield more readily to guidance by the grooves, less downward force is required to press the stylus against the groove walls; therefore, less wear is imposed on disc and stylus. In terms of tracking (groove adherence), high compliance allows the stylus to follow rapid, wide swings in the groove without, as it were, cutting curves and tearing up the road.

Compliance is stated in terms of force—as, for example, 20 x 10^{-6} cm/dyne. This may seem forbiddingly technical at first glance, but it simply means that if 1 dyne (a basic unit of force) pushes on the stylus, it moves a distance of 20 millionths of a centimeter. In comparing compliance ratings, the significant number is the one before the multiplication sign. The higher this figure, the higher the compliance. Since the specifications listed by different manufacturers are not always directly comparable, a more reliable measure of compliance is the minimum force at which the cartridge will track low-frequency signals. If it tracks well at less than 2 grams, it is excellent indeed. To operate at the minimum tracking force specified for high-compliance cartridges, the tone arm must be virtually friction-free. Cartridges with compliance much higher than about 15×10^{-6} cm/dyne can, therefore, be used only with top-grade arms.

A stylus assembly's dynamic mass can be loosely defined as the amount of weight the record groove has to push around in order to generate an electrical signal in the cartridge. The dynamic mass is not the same as the weight of the moving parts (diamond, stylus shank, etc.) because of the leverage effect of the cantilever design used in most stylus assemblies. Any mechanical device that has to stop and go at rates up to 40,000 times per second (to track a 20,000-Hz tone) must of course be light, the lighter the better. Otherwise, too much inertia develops and the stylus can't keep up with this fast shuttle. It either overshoots the curves or just cuts across them. Either form of mistracking is a musical calamity.

If, in an effort to lower the mass, the stylus shank is made too light, another problem arises: the very thin shaft becomes flexible and the motion of the diamond tip is not accurately transferred to the current-generating parts of the cartridge. Since the weight of the diamond is fixed by its dimensions and mounting, an optimum compromise must be worked out between weight and rigidity of the shank. Fortunately, modern metallurgy, stimulated by the requirements of space exploration, has developed some extremely tough lightweight metals. Taking advantage of these materials, cartridge designers have recently been able to reduce the dynamic mass of the stylus without losing high-frequency transmission along the shaft. The dynamic mass of a modern high-performance cartridge is usually specified at 1 milligram or less.

Stylus Dimensions

The mechanical behavior of a cartridge—as distinct from its electrical properties—is also determined by the shape and size of the diamond tip. Contrary to a widespread notion, the diamond is not sharpened to a conical point like a pencil. Such a point would rip the record no matter how light the tracking weight. Rather, the tip is rounded, and the radius of its curvatures varies among different cartridge models. Most manufacturers offer a choice of .7-mil, .5-mil, and .4-mil styli, 1 mil being equal to 1/1000th of an inch. Several manufacturers also offer 3-mil styli suitable for playing older 78's.

The smaller styli are capable of cleaner high-frequency reproduction because they fit more snugly into tight little curves, especially toward the center of a record where the musical waveforms are more densely packed. However, the .5- and .4-mil styli tend to rattle loosely in the wider grooves of some of the older monophonic records. The .7-mil stylus tracks both new and old records quite adequately and can be recommended as a universal stylus to track an LP record, mono or stereo, regardless of age.

The ideal of the phono-cartridge designer is a unit that yields to guidance from the record groove in a purely passive manner. But real-life cartridges invariably fall somewhat short of this ideal. Like all bodies acted upon by external forces, the moving parts of a cartridge are beset by inherent resonances that interact and interfere with the musical signals the cartridge must reproduce.

Cartridge Resonance

Trouble arises when the music or some of its overtones happen to hit the frequencies at which the moving parts of the cartridge resonate. The cartridge then makes its own special spurious contribution, and the resulting sound is often unpleasantly shrill. Violins, for example, mishandled by a resonance-ridden cartridge, seem to be made of stainless steel instead of wood.

Audio engineers refer to this type of sound as "peaky" because it is caused by frequency-response peaks (exaggerated response) at those points in the frequency spectrum where the music coincides with the cartridge's own resonance. These same resonant peaks cause the cartridge's channel separation to drop considerably. Since the sound structure of music covers a broad range of overtones, chances are that nearly every note contains some harmonic component that will excite some resonance within a poorly designed cartridge.

Cartridge designers have lately been successful in suppressing unwanted resonances or moving them out of the audible range by employing new materials and techniques to make the moving parts of the cartridge extremely light. The lighter a vibrating body, the higher (in frequency) its resonance, other factors being equal. Thanks to the lightness of their moving masses, the main resonance of the best modern cartridges has been pushed up beyond the audible range, that is to say beyond the limits of the recorded signal. As a result, modern top-quality cartridges are virtually free of spurious sound coloration and their stereo separation remains excellent over the frequency range of the recorded material.

Few manufacturers specifically state cartridge-stylus resonance as part of their specifications. In any case, the proof of this particular pudding lies in the uniformity (smoothness) of the cartridge's overall frequency response. Any listing of frequency-response limits (for example, 30 to 18,000 Hz) should be accompanied by a statement of how many decibels (db) the cartridge deviates from uniform response within the range. The deviation should be as small as possible. In the case of the top-price cartridges, the frequency-response deviation (above or below the signal-output level at 1000 Hz) should not exceed about 2 db. Square-wave test records are also valuable for detecting "ringing" and other resonance-related instabilities.

Some manufacturers include a frequency-response graph with their cartridges, and the test reports published in audio magazines often present such curves. The thing to watch for is an elevated portion of the curve covering a fairly broad part of the spectrum, particularly in the 8000-Hz to 15,000-Hz region. This is the mark by which resonance problems and a possible harshness of sound can be spotted.

Cartridge Theory

Most high-fidelity cartridges operate on the principle that an electric current can be generated in a coil of wire by varying the magnetic flux impinging on the coil. The power plants that light our cities produce electricity in this way, and one might think of a magnetic phono cartridge as a miniature electric generator. Relative motion between the magnetic-flux field and the coil produces a voltage that corresponds in frequency (pitch) and amplitude (loudness) to the musical content of the record groove.

It does not really matter whether the magnet moves and the coil remains stationary, or vice versa, or whether neither moves and some third element varies the magnetic interrelationship between the two. The engineer thus has the option of designing either a moving-coil, a moving-magnet, or what could be called a moving-flux cartridge.

Moving-magnet models are theoretically quite simple. The magnet is attached to the rear of the stylus shank and swings between two pairs of coils so positioned that one pair produces the left stereo signal while the other produces the right. Manufacturers who have settled on this type of phono cartridge have achieved excellent results. However, those who feel that a magnet swinging back and forth between the coils will not generate the most linear signal favor the moving-coil approach. With this technique, the coils move while the magnet stays put. Still another approach involves having a stationary magnet induce a magnetic field in the stylus shank whose motion is then magnetically sensed by the stationary coils or simply by having the shank "interfere" with the magnetic path between the magnet and the coil.

Magnetic cartridges are today's standard in the quality sound field and have proven their musical competence time and again. However, a few unorthodox designs on the market operate on different principles. In one of these, a light beam is deflected by tiny mirrors in keeping with the stylus motion in the groove and then converted into electric signals by a photoelectric receptor. But such esoteric designs are merely exceptions to the general rule of magnetic cartridges.

Another kind of cartridge, the so-called ceramic cartridge, is widely used in cheap, non-fi equipment. In these models, the stylus twists slabs of ceramic material which then produce a voltage proportional to the torsional force exerted by the stylus. Some of these produce a pretty fair signal. However, because the stylus must do relatively heavy mechanical work in twisting the ceramic slabs, it must push hard against the record groove. This hastens record wear and also makes it harder for the stylus to follow the rapid motion involved in high-frequency reproduction. For this reason, ceramic cartridges are not used in quality sound systems.

Equalization

In concluding this chapter on record reproduction, we should briefly explain the concept of equalization, although it is automatically taken care of by the amplifier circuitry and therefore of merely academic interest to the listener. However, if you have ever wondered how the full range of musical sounds gets squeezed within the narrow confines of the record groove, the answer lies in this process of equalization. Basically, it involves weakening the bass and strengthening the treble when the record groove is cut.

Why equalization is necessary becomes clear when you consider the physical nature of bass and treble sound. Low notes embody a great amount of sheer physical power. Recorded at their natural strength, notes from such instruments as a tuba, bass viol, or kettledrum would yield wide-swinging record grooves far too great in amplitude to be tracked by any known, or even feasible, phono cartridge. That is why the recording engineer applies bass equalization, which is analogous to looking at the sound wave through the wrong end of a telescope. The exact pattern of the sound wave is preserved, but its overall amplitude is reduced to make it produce a smaller groove. The lower the frequency of the note, the greater the degree of reduction.

Just the opposite problem exists in the treble range. Natural, unboosted high-frequency waveforms are so tiny that they would be lost in the minor imperfections of the record surface, much as small photographic details are sometimes lost in the grain of the film. Therefore, to prevent treble sounds from being buried in surface noise, the engineer strengthens the highs, making their waveforms bigger than lifesize so that they will stand out more clearly.

The net result of equalization is a disc with totally unnatural sound, weak in bass and strident in treble. To restore the natural balance, exactly the reverse of equalization must be applied in playback. The treble is cut back and the bass boosted. This is done automatically by the preamplifier section of the playback equipment. When the treble content of the music is reduced, so are the treble sounds that constitute surface noise. The effect is to minimize this noise in relation to the signal. What finally emerges from the speakers is, ideally, the whole sound spectrum restored to its natural balance.

In 1955, the Record Industry Association of America

(RIAA) adopted a standard equalization curve, specifying the precise amount of treble and bass equalization to be applied during playback. Since then, records and playback equipment have been made according to this standard in virtually all countries. When you switch to the magnetic phono input, the correct playback equalization is automatically applied. Records made before 1955 have one or another of several slightly different equalizations, but you can usually get fairly balanced sound from them by adjusting the treble and bass controls.

10 Tuners—Music from the Air

Writing during the Victorian era, Edward Bellamy blithely predicted that by the year 2000 all human strife would be ended. Taking his clue from the newly invented telephone, he envisioned music being conveyed by some electric means into every home. Thanks to universal exposure to such benign influence, he felt, all conflict would resolve itself in harmony.

Bellamy evidently underrated human resistance to benign influence, but his prediction has come true at least in the technical sense. With radio broadcasting, music has become a kind of natural resource, one of the less lethal additions to the atmosphere provided by modern technology. Anyone with a radio can, metaphorically, reach into the sky and haul down some of its musical bounty.

Our concern here is to maximize the quality of radio reception. People who have never heard radio programs through a component sound system are in for a surprise. Their initial reaction usually is, "But it doesn't sound like a radio at all!" In fact, it is possible to obtain stereo radio reception matching in tonal range and clarity the best records and tapes. For this you need a tuner, a component that pulls in the radio signals and feeds them to your amplifier. As we have pointed out before, some amplifiers and tuners are combined into a single unit, and what is said here in regard to tuners equally applies to these receivers.

FM AND AM

There are two methods of broadcasting: AM (Amplitude Modulation) and FM (Frequency Modulation). Only FM is capable of the full range of sound and of transmitting a stereo signal. AM, the older form of broadcasting, has the advantage of reaching greater distances, but it can hardly be considered a suitable medium for high-fidelity sound. It cuts off frequency response at about 5000 Hz and is generally beset by static and other sorts of noise. Many FM tuners, however, also offer additional AM reception for the simple reason that you may want to listen to broadcasts that are not available on FM. Even with its inherent limitations of tonal quality, AM will sound better than you ever heard it before when the signal is fed through your sound system. But for wide-range, high-quality sound, you have to stick with FM.

TUNER SENSITIVITY

Among FM tuner specifications, sensitivity usually gets top billing. Sensitivity measures the tuner's ability to pull in weak or distant stations. It is debatable whether very high sensitivity is really so important, except in fringe locations where reception is unusually difficult. But it is a safe assumption that those tuners (or receivers) that have been engineered for high sensitivity also excel in their other specifications.

Sensitivity is stated as a certain number of microvolts, indicating the minimum strength an incoming signal must have to accomplish a certain amount of "quieting" in the tuner. Quieting refers to the ability of an incoming signal to suppress the hissing noise normally heard on FM tuners between stations. The lower the number, the more sensitive the tuner.

Several years ago, the Institute of High Fidelity (IHF) proposed uniform standards for making sensitivity measurements. Suppose a tuner specification reads: "Sensitivity 3 μv IHF." This means that on this particular tuner an incoming signal of 3 millionths of a volt provides 30 db of quieting—that is, the distortion and noise level is 30 db below the audio signal. This isn't very quiet reception. It would take a somewhat stronger signal to make the background noise fade into complete silence. But as a reference standard, the IHF figure of 30 db is very useful, for it permits direct comparison of sensitivity ratings for any tuner tested under these conditions.

This top-quality tuner features a space-saving, drum-type dial and two separate tuning meters, one to indicate signal strength, the other for precise center-channel tuning. (*Photo: Harman-Kardon*)

Unfortunately, many radio manufacturers outside the high-fidelity component group do not observe IHF standards and publish sensitivity figures obtained under more lenient conditions. For example, a manufacturer may advertise the sensitivity of his set as "2 μv for 20 db quieting." Offhand, this would seem superior to a tuner with, say, 4 μv sensitivity. But if you look at the numbers closely, you notice that the seemingly more sensitive unit provides only 20 db quieting at the stated signal strength, while the IHF testing standard calls for 30 db. Sometimes nonhi-fi manufacturers put the quieting figure in small print or omit it altogether, deliberately rendering their specifications misleading or meaningless. In comparing tuner data make sure, therefore, that the statement of sensitivity includes the letters IHF. In virtually all measurements, they represent a sort of badge of reliability.

Actually, it would be more meaningful to rate tuner or receiver sensitivity at a higher level of quieting—say, 50 or 60 db. For in practice, signals with only 30 db quieting are still marred by background noise. Some tuner spec sheets therefore give a figure for "full-quieting" sensitivity—the minimum signal strength which will allow the tuner to provide maximum quieting.

Just how much tuner sensitivity do you need? This depends on the location of your home and the surrounding terrain. As a general rule, an IHF rating of 2 to 4 μv should provide satisfactory reception up to a distance of fifty or sixty miles from the transmitter, and, with the aid of a roof antenna, even beyond that. Higher sensitivity ratings between 1 and 2 μv are needed only in extreme fringe areas.

AM REJECTION, SELECTIVITY, AND CAPTURE RATIO

But sensitivity is not the only significant factor in estimating a tuner's total merit. In most critical reception areas, where the incoming signals have ample strength, satisfactory tuner performance hinges on other qualities which, oddly enough, are rarely considered by the prospective buyer. Among them: *AM rejection, selectivity,* and *capture ratio.*

AM rejection describes the tuner's ability to squelch electrical interference, whether it is atmospheric static or man-made electrical noise. Many manufacturers now list AM rejection among the FM tuner specs. An AM rejection rating of −45 db or better will give satisfactory results in most situations, but figures of −50 or −60 db are increasingly common on top-quality tuners. AM rejection is also very important if you live in an area plagued by *multipath* reception problems. These multipath problems result from the same kind of signal reflections that produce "ghosts" on TV screens. In FM reception this can cause a great deal of distortion, especially in stereo. Some authorities even consider AM rejection the most important FM tuner specification, especially in urban areas, where signals bounce off steel-frame buildings, causing many forms of multipath distortion.

Selectivity, often overlooked in tuner specifications, can be important in both urban and rural locations. Selectivity refers to the tuner's ability to keep separate any stations that are close together on the dial. If you live in areas where only a small number of FM stations are within receiving range and they happen to be widely spaced out across the dial, selectivity is obviously not a problem. But in some metropolitan and suburban regions the FM band is getting so crowded that selectivity is essential. A tuner with a selectivity of 45 db or more will spare you the annoyance of overlapping signals, and tuners with up to 100 db of selectivity are now available.

Cross-modulation rejection refers to the tuner's ability to accommodate a wide range of signal strengths. In nontechnical language, it describes the tuner's resistance to being "overpowered" by a strong signal from a nearby station. If this happens, the tuner's input stage "cross-modulates"; *i.e.,* the strong station appears at several different points on the dial, often blanking out other stations. A tuner with a cross-modulation rejection of 80 db or thereabouts is not likely to run into that kind of trouble even if you live within a stone's throw of the transmitter. Moreover, recent improvements in transistor antenna-input circuits and the use of field-effect transistors (FET) make tuners better able to cope with excessively strong signals.

Cross-modulation is actually a form of *spurious response,* and on some spec sheets cross-modulation rejection is therefore called *spurious-response rejection. Image rejection* is another form of spurious-response rejection. Image rejection chiefly defines a tuner's ability to reject signals in the aircraft communications band. Unless you live near an airport, it may be of little importance to you.

Capture ratio is a term that describes the tuner's ability to separate stations broadcasting on the same frequency. If two stations come in on the same spot on the dial, the tuner should "capture" the stronger one clearly and suppress the weaker one. The capture ratio tells how much stronger one of the two conflicting signals must be to bring about this complete separation. Hence, the lower the number, the better the performance. For example, a capture ratio of 4 db is considered adequate, and a ratio of 2 db is excellent. (It means that one station need be only 2 db stronger than the other to override interference from it.)

Capture ratio used to be important only to FM listeners living between two cities each of which had FM stations operating on the same frequency and reaching the listener's location with approximately the same strength. But now, with stereo, capture ratio performs an added function. It helps reject multipath signals, which—as we have explained—correspond to ghost images in TV. Good capture ratio thus contributes to the clarity of stereo reception and to the maintenance of proper stereo separation, unencumbered by the vagaries of signal reflection from the surrounding terrain.

To remember the difference between capture ratio and selectivity, remember that selectivity refers to the ability of the tuner to separate stations that are *adjacent* on the dial, while capture ratio refers to the tuner's ability to suppress one of two stations that fall at the *same* place on the dial.

DISTORTION AND S/N RATIO

One tuner characteristic—distortion—is often overlooked. A total harmonic distortion of about 1 percent is quite respectable, and topnotch tuners will operate at less than .5 percent distortion when measured at 100 percent modulation—that is, at full tuner output.

Signal-to-noise ratio (S/N) is measured on tuners just as it is on amplifiers. But on FM sections, the signal-to-noise ratio also tells you the tuner's ultimate quieting capability.

ANTENNA AND CONTROLS

It has often been said that no tuner is any better than the antenna to which it is attached. Certainly a good antenna gives your tuner a far better chance of catching FM stations clearly and reliably. The farther you are from the station you're trying to receive, the better your antenna should be. Moreover, a good antenna improves the performance of any tuner, regardless of its sensitivity. If you live in a city where all the stations are close by, an indoor antenna may suffice. If you're in the suburbs, you may want to add a roof antenna for best reception, especially for stereo where a good strong signal contributes much to the quality of sound. If you live in an apartment house with a master TV antenna, try plugging your FM tuner into the TV antenna socket. Sometimes this works quite well. And in fringe areas or low-lying locations, you may need a multi-element antenna and possibly a rotator similar to those used for TV.

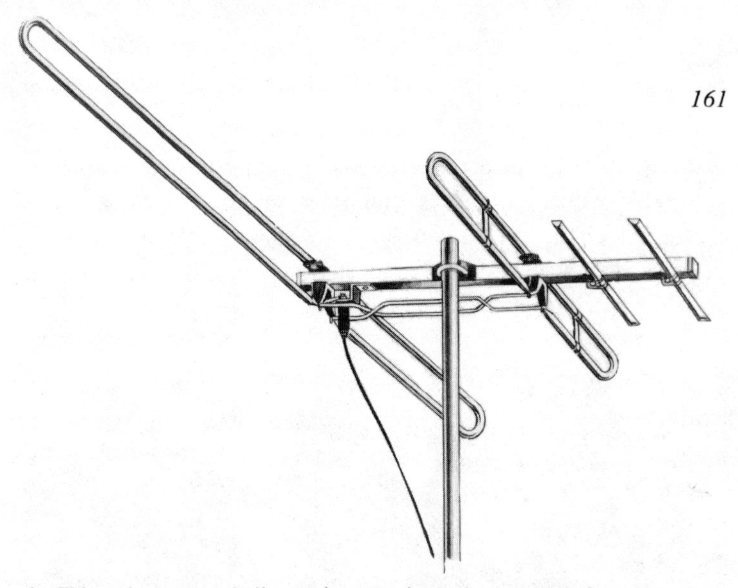

An FM roof antenna similar to the one shown here greatly improves reception in many areas and is a necessity in fringe locations. (*Jerrold Corp.*)

As for tuner controls, all you really need is some way of selecting the station you want to hear. In most cases this is done by means of an ordinary tuning knob, just as on a regular radio. But unless an FM station is very accurately tuned in, the resulting sound will be fuzzy and distorted. For this reason, virtually all better tuners have some visual indicator to show when the station is tuned correctly. These tuning meters come in two types. One simply swings from left to right when a station comes in, and you adjust your tuning for maximum swing of the pointer. What this meter measures is the strength of the incoming signal. That's why you can also use it to aim your antenna for maximum signal pickup if you have a rotating roof antenna. A far better type of tuning meter is the "center channel" type. You adjust your tuning until the pointer is exactly at a marked spot in the middle of the dial. This shows that the frequency of the transmitter is exactly aligned to the frequency selected on your tuner. Some deluxe tuners have both types of meters.

If you don't want to bother with the necessary niceties of accurate tuning, you can get—for a price—automatic tuning features. A few digital tuners are on the market now on which you select your stations as you select a number on a push-button telephone. For example, for a station with a frequency of 96.3 Hz you'd just push buttons 9–6–3. Instead of a dial, such tuners simply have a panel on which the selected numbers

A tuner with digital read-out assures easy accurate tuning, but this convenience adds considerably to cost without inherently improving sound quality. (*Photo: H. H. Scott, Inc.*)

light up. Other automatic tuners feature a "hunt-and-seek" mechanism that automatically scans the dial while you push a button. As you release the button, the automatic mechanism locks to the nearest station with complete tuning accuracy. Still others have preset tuning for FM stations, similar to the button tuning on car radios. These, however, tend to be somewhat less accurate and often require resetting.

All these automated tuning features tend to be rather expensive and, in themselves, add nothing to the quality of reception. So, if you are not too lazy to use the old-fashioned tuning knob for changing from one FM station to another, you can save yourself a lot of money and invest it where it really counts toward greater sound quality.

Some tuners have a switch marked AFC, which stands for Automatic Frequency Control. This circuit clamps a sort of hammerlock on the incoming FM signal, keeping it accurately tuned in. However, AFC has a tendency to latch on to strong stations and cover up the weaker ones. Therefore, to tune in a weak station, it is necessary to switch off the AFC. Modern tuners are so improved in their circuits that stations rarely drift out of tune and AFC is not really necessary. Some designers consider it more bother than it's worth. So if you don't find an AFC switch on a tuner, this doesn't reflect on its quality.

Many tuners offer a muting switch, which blocks out the noise that appears between FM stations as you tune from one to another. However, such muting is attained only at a loss of sensitivity to weak signals. If you are only listening to strong, nearby stations, leave the muting switch on. To pull more distant stations, you may have to turn the muting off and put up with a little interstation noise.

Most tuners have a signal light that automatically lights up when the station to which you are tuned is broadcasting in stereo. All but the least expensive tuners have an automatic switch that will switch off the stereo circuits when a mono broadcast is tuned in. Many tuners also have a manual switch for changing from mono to stereo. If a stereo station won't come in clearly but keeps on sounding fuzzy no matter how carefully you tune it, you can usually clear up the sound by throwing this switch to mono.

Some of the latest tuner designs already have plug-in provisions for adapters that will permit reception of four-channel sound if and when the FCC ever approves such broadcasts. If you are future-oriented and seriously thinking of "going quad" this may be a consideration in your choice.

11
Tape Recorders—Reel, Cartridge, and Cassette

During the later phases of World War II, the overnight disappearance of their cities persuaded many Germans that things were not going according to plan. At this point, Adolf Hitler felt that his people needed comfort and encouragement, best provided by steady exposure to his reassuring voice. His recorded harangues were constantly replayed on all radio stations, blaring in the streets and squares from the public-address loudspeakers installed on virtually every lamppost. Still Hitler wasn't happy. The disc recordings of his day didn't satisfy him. Their surface scratch and limited frequency range cut short his charisma. Besides, they were cumbersome to record. So, Dr. Goebbels, the propaganda chief, ordered a crash program for the development of more convenient and convincing recording techniques. The result was the invention of magnetic tape recording.

No longer linked to Hitler's megalomania, tape recording has become one of the most versatile means of sound storage, and this versatility, plus ease of use, makes the tape recorder a valuable adjunct to any audio system.

To begin with, a tape recorder is the perfect music-stealing machine. Leaving aside whatever ethical questions may be involved, you can simply tape music off the air—free. And if you tire of a piece, you can use the same tape again to record another selection. Or you might copy your friends' records onto tape, perhaps obtaining performances no longer available in the shops. Or you might want to tape irreplaceable items in your own record collection so that you preserve the music even after the record is worn out. For tape is immune to the scratches and mechanical erosion that limit the life of records. Some kinds of tapes, the mylar-based types, last practically forever. In fact, they are among the most durable materials known.

Aside from bringing you musical enjoyment, a tape recorder lets you document your personal history in sound—the speech of your children at various stages of their growing up, a family reunion, your latest sales presentation, or your speech to the local action committee.

Many different kinds of tape recorders are available today, and your intended use will determine the choice. The main options are: portable or home-based. In either category, you also have the choice among three different tape systems: open-reel, cassette, and cartridge. In basic principle, all three are alike. As the tape rolls past the recording head, it gets magnetized. The amount of magnetization varies exactly as the electric signal representing the sound. A magnetic replica of the sound waves is thus created. In playback, the process is reversed. As the magnetized tape rolls past the playback head its magnetic impulses generate electric signals which are then played through the loudspeaker. The recorder also has a third head, which can demagnetize the tape and thus erase previously recorded material. This allows you to use the same tape over and over again.

Four separate tracks are recorded across the width of the tape, two in each direction. The two tracks represent the two stereo channels. At the end of the reel, you flip over the tape and play the two other tracks in the other direction. On open-reel recorders you also have the option of recording each track separately, in mono, which doubles the available playing time per reel. This is an economy factor in speech recording, where stereo isn't important.

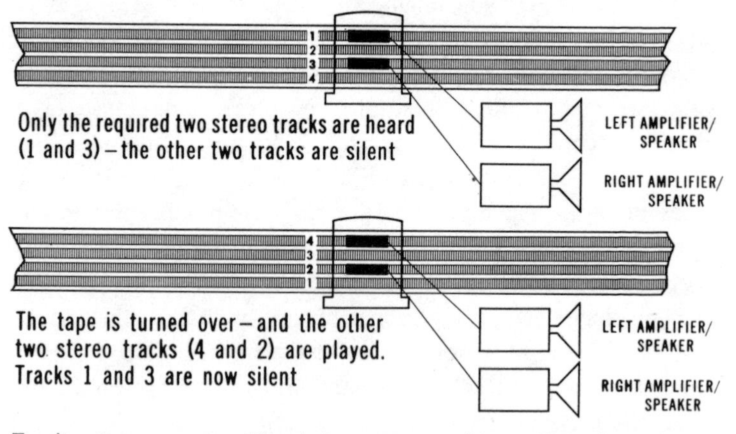

Track pattern on conventional stereo open-reel tape. Four channels are recorded side by side across the width of the tape. Channels 1 and 3 are played in one direction, channels 2 and 4 in the other direction after the reel is flipped over.

Open-reel recorders still offer the best fidelity, and if that is your prime consideration, an open-reel machine is your obvious choice. However, the best among the current cassette machines—the cassette decks designed specifically for use with high-quality sound systems—closely approach the sound quality of open-reel machines and offer unique advantages of compactness and simplicity of operation. In a cassette machine, all you do is insert the cassette, push the button, and you're ready to roll. In an open-reel machine you must thread the tape through the recorder. This isn't difficult by any means, but if you are fumble-fingered you may prefer the convenience of the cassette.

We have so far mentioned two of the three current tape formats, open reel and cassette. The third, the cartridge tape, is used chiefly in car stereo systems. These cartridges do not match the better cassettes in convenience or in fidelity, and it is probably a fairly safe prediction that, as car manufacturers are beginning to switch from cartridge to cassette even for automotive stereo, the cartridge will fade from the field and leave the cassette as the dominant medium for magnetic recording. Since cassette and cartridge equipment are not compatible, this would be a welcome step toward standardization, even though it would make all cartridge recorders and players obsolete. Meanwhile, the cartridge, with its cumbersome eight-track system, is best left to those who listen to music mostly while driving in a noisy car, where fidelity is not a prime consideration. For serious home listening, we need concern ourselves only with open-reel and cassettes.

The fidelity of a tape recording is partly a function of the speed with which the tape travels past the recording and playback heads. The reason for this is that at slower speeds, the magnetic impulses imparted to the tape get crowded closer together. When they come in very rapid sequence, such as at high frequencies, they may overlap and thereby become obscured. But faster tape speed, while improving fidelity, shortens playing time per reel. Also, it uses more tape for a recording of a certain duration and thus adds to the tape cost.

To let you pick the optimum compromise between tape speed, playing time, and economy, virtually all open-reel recorders allow you a choice of different tape speeds. The usual options are 7½ inches per second (abbreviated ips), 3¾ ips, and 1⅞ ips. A few machines even offer an extra-slow speed of $^{15}/_{16}$ ips. At 7½ ips a good tape recorder yields a frequency response more than ample for even the most exacting requirements. The 3¾ ips speed still provides highly satisfactory musical fidelity on a good machine, but 1⅞ ips, because of its somewhat re-

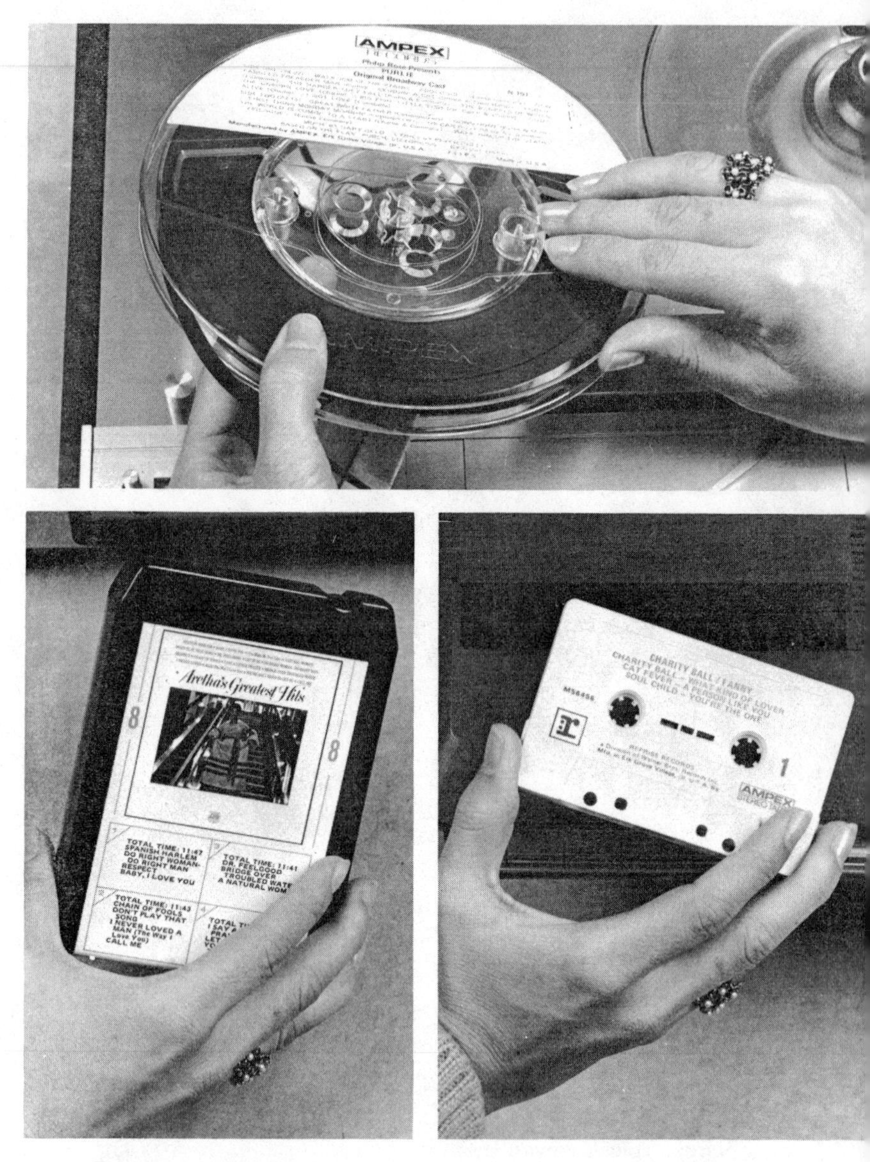

The three basic tape formats: open-reel, cartridge (lower left), and cassette (lower right). Only open-reel and cassette can be seriously considered for high fidelity.

stricted range, is intended for speech recording only. The same is true of the even slower 15⁄16 ips speed recently introduced on some machines. At 7½ ips, a normal 1200-foot tape reel yields one hour playing time in stereo, two hours at 3¾, four hours at 1⅞, and eight hours at 15⁄16. For monophonic operation, the playing time is doubled in each case, since only one track is used at a time. Typical frequency response figures for a good recorder at various speeds might read 30–18,000 Hz at 7½ ips, 30–12,000 Hz at 3¾ ips, and 50–7000 Hz at 1⅞ ips.

OPEN-REEL TAPE DECKS

For most home installations where high fidelity is the main object, an open-reel tape deck is the logical choice. A tape deck is a tape recorder minus amplifiers and speakers. Instead, the tape deck plays through your amplifier and loudspeakers, functioning as an integral part of your home music system. Like standard tape machines, a tape deck lets you record either live via microphone, or directly from FM tuner, TV, or records. Connections between the deck and your TV set and the rest of your sound system remain permanently set up and neatly hidden from view.

If fidelity is your chief aim, you'll want a deck with a top speed of 7½ ips. Your deck should have some lower speeds as well: 3¾ ips is a necessity; 1⅞ ips is handy for economy speech recording, allowing you to cram lots of material on a single reel.

Open-reel decks are available with one, two, or three motors. The three-motor units are the most reliable because they don't rely on a lot of pulleys, belts, idlers and linkages to get the rotating power where it's needed. They also rewind the tape faster. But they cost more than single-motor machines. Two-motor decks rank midway between those of the other two types. Three-motor decks sometimes have relay-operated controls. Like some elevator call buttons, these operate at the slightest touch. This is a convenience, nothing more, and an expensive one. Other machines have the same control features, but you have to turn levers instead of just touching those responsive buttons. But the soft touch of the buttons and the smooth clicking of the relays may appeal to the Walter Mitty in you. Some relay-controlled machines offer optional remote control, so you can operate the recorder from your armchair.

More and more tape decks incorporate reversing switches, to play through both sides of a four-track stereo tape without your having to flip reels. On some units this reversal is automatic at the end of the reel. However, in most cases the automatic reverse only operates in playback, not in recording.

A standard open-reel tape deck—this one featuring three speeds and three heads. (*Photo: Allied Radio Shack*)

Some decks have line inputs (for signals from the tape outputs of your sound system) and microphone inputs with individual volume controls. This lets you mix live voices with recorded sound effects or music, for dramatic tapes, slide-show and home-movie soundtracks, and the like. On most decks, though, there's only one recording level control; it usually controls the line input, until you plug a microphone in.

A deck with separate heads for erase, record, and playback has two advantages. For one thing, heads built specifically for recording or playback are better suited to their job than compromise heads designed for both functions. Besides, a separate playback head lets you monitor the quality of the signal on the tape while you record it. That way, you can immediately correct the setting of the controls if the recorded signal isn't quite up to par.

CASSETTE DECKS

When utmost fidelity is not the overriding consideration, a cassette deck offers advantages of simplicity and compactness unmatched by any other type of recording equipment. Like the open-reel decks, cassette decks are designed to operate through your regular amplifier and speakers. But they are considerably smaller than reel decks, and hence fit more easily on a shelf. Besides, their remarkable ease of operation makes them attractive to people who don't want to fumble with loose ends of tape.

I have directly compared the fidelity of some top-rank open-reel decks with some top-rank cassette decks. I recorded the same music simultaneously on both machines and then compared the result in playback. I am almost tempted to say that I couldn't tell the difference. But such a statement would reflect my surprise, pleasure, and enthusiasm rather than the truth. When I concentrated my attention on such details as the definition and clarity of the uppermost highs, the transparency of orchestral texture or the open sheen of a voice, I could discern a margin of

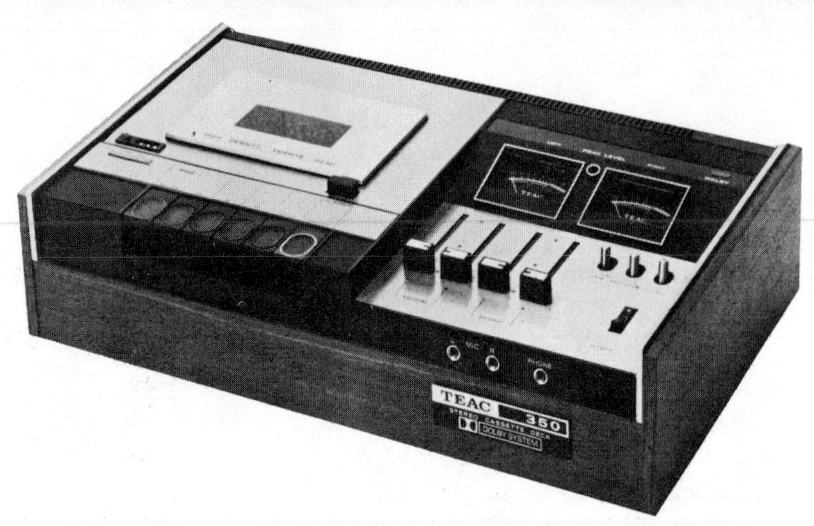

A top-quality cassette deck with built-in Dolby noise suppressor. (*Photo: TEAC Corp.*)

quality in favor of open-reel tapes. What surprised me was the smallness of the difference, evidence of the great advances in cassette technology.

Granted, that's a subjective evaluation. But in some ways it may be more informative than comparing engineering data. Perception of music, after all, is a subjective matter. To put it

Cassettes can also be used for high-quality car stereo, as with this automotive stereo cassette player. The same cassettes can then also be used in a home system.

another way: I have no reservations, either musical or technical, in recommending one of the better cassette decks even to a finicky music lover if he feels that compactness and simplicity of use are important criteria of choice.

PORTABLE RECORDERS

The compactness of the cassette format has led to the development of some excellent, battery-powered portable recorders. Naturally, these camera-sized models cannot compare with the larger, heavier, home-based cassette decks, but I have found them handy in making on-location recordings at jam sessions and in nightclubs. In playback, of course, they are limited by their puny built-in amplifiers and speakers. I generally use the portable machine for recording only, then play back the cassettes through my large home-based system.

If you are a footloose type and want to take your recorder to all sorts of unlikely places, a battery-powered portable is your obvious choice. Unlike any other kind of recorder, it won't tie you to the nearest wall plug. You can go on tape safaris, bagging your material wherever you find it—on street corners, in taxicabs, weekend hideouts and hunting lodges, sports events, folksong hootenannies, hay rides, nightclubs, church choir rehearsals, or the initiation rites of the Navahos. Some portables are so small that you can slip them under your coat in a holster, like a gun. The microphone is either built in, or you can get a microphone disguised as a wristwatch. You've got the cable up your sleeve, safely out of sight, and can gumshoe around for some sneaky electronic eavesdropping.

A word of caution. Steer clear of transistor tape machines sold for $50 or less in glorified junkshops and other bargain emporia. Such machines are often cheaply built, erratic in operation, and generally afflicted with "flutter and wow"—the bugaboo of poor recorders that makes all music sound tremulous and shaky.

To get a good recording, you must make sure that the signal on the tape neither drops low enough to be swamped in background noise nor rises high enough to cause distortion. Most decks and recorders make you adjust levels yourself, while checking recording levels on a meter. Most cassette portables, though, have a feature known as ALC (Automatic Level Control). This automatically evens out the recording level, so you don't have to watch the needle on the recording meter and adjust the gain control. Loud, soft, near or far sounds all get on

Cassettes permit portable recorders to be so compact that they can be carried in one hand. Playback quality greatly improves when recordings made in the field with such equipment are played through larger home sound systems. (*Photo: 3M/Wollensak*)

the tape with about equal loudness. That's dandy for recording talk. But stay clear of it for recording music. Since it makes all sound come out equally loud, it kills most musical phrasing and expression.

Some recorders, therefore, let you switch off the ALC and control recording level manually when you tape music; a few others have two ALC positions—a fast-acting one for speech and a more gentle one for musical recording.

Batteries can get expensive. A portable that can be played on house current when you're near an outlet will prove less expensive in the long run than one that works only on batteries.

Most small-size portables record in mono only. Stereo portables are available, but they are bigger, heavier, and more expensive.

RECORDER SPECIFICATIONS

Many tape recorder specifications are similar to the amplifier and speaker specifications (frequency response, distortion, and so forth) discussed earlier. Mechanical factors describing speed constancy also enter into the picture, notably wow and flutter, and these have already been dealt with in connection with phonograph turntables.

Unfortunately, these familiar expressions are usually applied more loosely for tape recorders than for other components. Frequency response, for example, is too often specified without a qualifying "plus-and-minus" number showing the decibel deviation from flat response. Such disregard for the essential matter of uniformity in frequency response sometimes leads to puzzlement and absurdities, such as when cheap recorders sport more impressive response figures than expensive, conservatively rated machines. If a manufacturer states the deviation at all, that fact alone lends a certain trustworthiness to his claims.

A statement of frequency response in tape recorders is usually based on the entire record-playback cycle. Occasionally, one finds that the frequency-response figures given refer to playback only. This describes the machine's playback frequency performance on prerecorded tapes, and in general, if a recorder does well when playing tapes, it should do nearly as well in the record mode.

The signal-to-noise ratio, an important factor in tape recorders, should also be read warily. The figure expresses the difference in decibels between the noise imparted by the recorder to the tape and the loudness of a recorded test tone. Although the National Association of Broadcasters has established a standard reference level for this tone, audio tape-recorder manufacturers do not all observe this standard. Some use a much higher reference-signal level (with up to 5 percent distortion), which yields a seemingly better signal-to-noise ratio. Again, the end result of such practices is that some inferior machines appear—on paper—to outperform their betters.

Under such anarchic conditions, unless the NAB (National Association of Broadcasters) standard is used, no direct comparison of noise figures is possible. In the absence of standards, the buyer's best bet is, if possible, to compare tape recorders by listening tests in a quiet environment. For example, check for the amount of hiss added to a tape recording of a disc, compared with the original disc sound. Of course, the machine should be hooked up to a wide-range audio system and high-quality tape should be used.

Similarly, distortion figures given for tape recorders are largely meaningless because of inadequately stated measurement standards. Even if the manufacturer notes that his distortion measurements are taken at "maximum" recording level (and "maximum" has no standard definition itself), other vital factors are not considered, such as the kind of tape used. Again, careful listening tests can help supply missing performance information.

Fortunately, the specifications referring to constancy of tape speed are more straightforward and can be directly compared. Expressed as a percentage, wow and flutter should not exceed .2 percent in a high-quality recorder.

SHOPPING HINTS

There are a few tricks to shopping for a tape recorder. The first is not to be cowed by the salesman or any technical gobbledygook he might spout at you. Even if you don't know a watt from a decibel, there's a fast, practical way to size up the performance of any recorder. Just put it through its paces. Work the various controls, observe if the tape starts and stops smoothly, without jerking. Check if the machine runs quietly, without vibration or whirring. Switch back and forth between REWIND and FAST FORWARD. This will reveal any possible tendency to break or spill tape, disastrous events if they happen in the middle of a recording session. A good machine treats the tape gently. But if a cassette machine binds or jerks, try it with another tape cassette. Sometimes the jam is in the cassette rather than in the recorder.

For a test that really puts a recorder on its mettle, bring one of your own phonograph records to the shop. Make sure the record is unscratched, dust-free, and otherwise properly pristine. Then ask the salesman to dub the music from your record onto the tape recorder you want to evaluate. Use direct cable connections rather than the microphone. In playback, alternately switch from record to tape (playing through the same speaker) and see how closely the tape copy duplicates the original sound on the disc. Pick music with long, sustained notes (piano tones are best). If they waver in pitch when you play them on the tape, it's a sure sign of wow and flutter.

A voice recording is a good test, especially for portable recorders. But never use your own voice for this test. Your own voice always sounds unnatural to you on recordings. (Reason: While you speak, you hear yourself by bone conduction from mouth to ear through your head. When listening to a recording of your voice, you hear your own voice as you would somebody else's.) Hold the microphone about eight or ten inches from the person speaking; closer, and you'll get exaggerated bass response, distortion, and unnaturally loud lip smacks, tooth clicks, and gurgles. Check playback for clarity of reproduction, especially for crispness of such consonants as the letters R and T, and S sounds that are clean, not hissy.

If the salesman won't let you make these tests, thank him politely and go elsewhere. Most specialized audio dealers will gladly set up such a demonstration, providing you don't barge in at the height of the Christmas rush. You can't very well expect the salesman to spend a leisurely hour with you if he's losing three sales in the meantime. But if you buy a pig in the poke, it'll squeal—especially if it's a tape recorder.

The Dolby

Many of the better cassette recorders, and a few of the open-reel machines, feature a Dolby noise suppressor. Basically the Dolby is an adjunct to tape recorders that reduces the hissing background noise inherent in the tape itself. However slight, this background noise is often noticeable during very soft musical passages, and recording engineers have long sought ways to eliminate it.

Earlier methods for reducing background noise on recordings depended on squeezing the dynamic range, i.e., shortening the span between the softest and loudest sounds on the tape or the record. During soft passages of the music, when background noise becomes obtrusive, the recording engineer used to crank up the volume slightly so that the music would override the background noise. But in doing so, the engineer frustrated the artist, and many a delicate, floating pianissimo was thus turned into a sturdy mezzoforte. The Dolby dehorns this dilemma. By reducing background noise it permits softer sounds to be recorded at their natural level. Not only does this keep soft sounds in character but it also retains more of the difference between extremes of loud and soft than was previously possible.

The technical principle of the Dolby is rather intricate. Two types of Dolby systems are available, one for home use and one for the professional recording studio. The studio device slices the total range of musical sounds into four segments—something like soprano, alto, tenor, and bass. Each range is scanned separately by electronic monitors and if the signal in one range falls below a certain loudness, it is automatically boosted. Later, during playback, it is cut back exactly to its original level. This cuts back noise by the same amount so that it becomes inaudible while the musical signal emerges at its natural level. The home system functions on the same principle, but operates on only the high-frequency portion of the signal, where such defects as tape hiss are most apparent.

Since this corrective action takes place only in the frequency segments where noise becomes a problem because of low signal level, the rest of the frequency spectrum passes undisturbed. This selective action partly accounts for the Dolby's musical merits, for it enables the Dolby to operate wholly without musically objectionable side effects.

The Dolby process works correctly only when it's used in both recording and playback of a tape or record. An ordinary tape played through a Dolby circuit will sound flat and drab, without high frequencies; a Dolbyized tape played without the Dolby circuit will sound shrill. Commercial cassette recordings, though, are often issued in their Dolbyized state. Played back through a home Dolby unit, they sound terrific. But played back without this circuit, they'll require a cutback of your treble control settings to sound reasonably natural.

Is the Dolby worth the extra cost? If you are a perfectionist, if your stereo system is good enough to really show the difference, and if you insist on cassette performance comparable to open-reel tape—then the answer is yes. If not, you can get good performance from a non-Dolby deck and save yourself a hunk of cash.

TAPE TYPES

Even the best tape machine is only as good as the tape you feed it. The improvement in performance resulting from premium tape is especially notable in cassette machines, where the slow speed of tape travel (1⅞ inches per second) makes particular demands on the magnetic coating on the tape. I occasionally use inexpensive bargain cassettes to record talk shows off the air where fidelity doesn't really matter. Unfortunately, these bargain cassettes sometimes jam up because of mechanical imperfections—in which case they cease to be a bargain. I never use such cassettes for music recording. For maximum performance on cassettes, you may prefer the special tape formulations known variously as Ultra-Dynamic or Super-Dynamic cassettes—or words to that effect. If your machine is adjustable to their special requirements, you may like the particular brilliance of sound obtainable from cassettes using chromium dioxide as magnetic material instead of the usual iron oxide.

Cassettes come in several formats, according to the length of time they play. They are designated by a number indicating total playing time in minutes. Thus, a C-60 cassette plays 30 minutes on each side, a C-90 cassette plays 45 minutes per side, and a C-120 plays 60 minutes on each side. Because the C-120s use thinner tape (to cram enough tape for two hours' playing time on the tiny cassette hubs) they are more likely to break or jam than the C-60s or C-90s. Avoid them, unless you really must have 60 minutes of uninterrupted recording time per side.

For open-reel machines you have a bewildering variety of tape types to choose from, and finding the optimum kind for your particular use may take a bit of experimenting.

Recording tape is a strip of plastic with magnetic particles stuck on. The properties of the tape depend largely on what kind of plastic is used as a base. The choice is between two main types: acetate and polyester. The latter, a film version of Dacron textile fiber, is better known by the DuPont trade name of

Mylar. Don't ask which of the two is better. That's likely to start arguments. But here's how they compare.

Mylar and Acetate

Mylar is twice as strong as acetate, but it has the nasty habit of stretching like taffy when pulled. If that happens, anything recorded on the stretched part is ruined beyond repair. By contrast, acetate breaks clean, almost without stretching. You can splice the broken ends without loss of recorded material. You can also buy tapes of special stretch-resistant Mylar, known as tensilized or prestressed tapes. It takes quite a tug to pull them out of shape, but many professionals still don't trust it. "I'd rather risk an acetate break I can patch than a Mylar stretch that would ruin an irreplaceable master tape," says a top studio engineer. But unless you do professional recording, chances are you're better off with Mylar. It's no fun to have to assemble your tapes from broken bits and pieces.

Mylar costs a bit more, but it lasts a lot longer. Acetate dries out and gets brittle. After fifteen years or so, it may fall to pieces. Nobody knows as yet how long Mylar lasts. Chemically it's more durable than steel or stone. So if you are recording for posterity, a Mylar tape may be better than a carved stone for durability. There's an easy way to tell tape bases apart: hold the tape reel or cassette to the light. If you can see light through the edges of the tape, it's acetate; Mylar is opaque.

Running Time

Then there's the question of running time. The thinner the tape, the more of it can fit on a reel. The standard tape thickness is 1.5 mil (1 mil = 1⁄1000 inch). A 7-inch reel of such tape runs half an hour at 7½ ips, one hour at 3¾ ips. But you can get thinner versions that cram two or three times as much tape on a reel and run two or three times longer. Commercially, such tapes are known as "extra play" (1½ times longer), "double play," and "triple play" tapes. You can even get quadruple play tapes, but they are frail and limited in bass response.

Tape Bargains?

Top brand names are priced pretty much in the same range for the same kind of tape. But if you are on a tight budget, try some of the house brands offered below standard prices by some of the larger electronics mail-order supply companies, such as Allied Radio or Lafayette. Chances are that you'll get good results—most of the time.

The main difference between top brands and house brands, aside from price, is quality control. The effort of making every inch of tape live up to specs is what accounts for the higher cost of the top brands, such as TDK, Maxell, and 3M Scotch Tape.

The makers of house-brand tapes are more relaxed about inspection and tolerances. That's how they can produce more cheaply. You may find that recording characteristics vary from reel to reel. Many will be fine. Occasionally you're apt to hit one with muffled highs, tinny bass, or persistent hiss, but considering the price, maybe you can take the chance of an occasional dud.

Some bargain tapes may be no bargain at all. These are called "white box" tapes because they come in unmarked boxes. At discount houses, you may be able to buy five of these for the price of one top-brand tape. But some audio fans call them junk tape. If you're lucky, you're getting some reject computer tape in those white boxes. It's designed for very high frequencies—about two hundred times higher than the highest sounds—to record the rapid pulse sequence of computers. So bass response is poor. But if you're only recording speech or don't care much for kettledrums and tubas, the results can be pretty fair. Occasionally the white box contains reject TV tape. That's another story. Unlike audio tape, video tape has all the magnetic particles standing upright instead of lying down sideways. As a result, such tape will hiss at you like a steam kettle.

Since TV tape is two inches wide, it must be slit down to the quarter-inch width of audio tape before being sold to hi-fiers. The reject mills are usually quite sloppy about slitting tolerances. The uneven cut makes the tape weave up and down as it travels across the recording head and the signals fade in and out like on a seasick radio. Besides, the coating on TV tape is too thin. It gets magnetically saturated by strong signals and distorts their sound.

Worse yet, the oxide layer on these rejects may have tiny bumps, dips, and holes. When a bump comes along, the highs suddenly disappear. A dip in the oxide layer makes the bass drop out. A hole makes everything drop out. So, if you want to be sure of a good recording, stick with standard brand tapes.

MICROPHONES

In making live recordings—whether on reel or cassette—the microphone you use has a decisive effect on the results you get.

In the typical home recording situation, it is the microphone rather than the recorder which sets the limit on attainable fidelity. Consequently, the majority of home recording fans have much to gain from upgrading the quality of their microphones.

Microphones furnished with low-priced tape recorders by the manufacturer are often of rather poor quality. Intended mainly for speech recording, they lack the range and refinement required for lifelike musical results. Makers of professional-type tape recorders rarely supply microphones with their products, leaving the choice to the user.

Each of the three basic types of microphones—ceramic, dynamic, and condenser—has its specific range of applications. Crystal or ceramic mikes are cheap, sturdy, and limited in fidelity. They are adequate for public-address or paging systems. Condenser mikes are usually found only in professional studios, although expert amateurs with top-grade recorders may be willing to spend upward of a hundred dollars for such a microphone to assure themselves of optimum results. Most amateur recordists, however, will find dynamic microphones in the price range from about $40 to $100 best suited to their needs. As a group, these microphones are capable of making surprisingly lifelike recordings on almost any of the better home-type tape recorders. Some of the less expensive condenser or "electret" microphones available in this price range give musically excellent results.

To match the microphone to the tape recorder it is necessary to consider its impedance—an electrical factor expressed in ohms and symbolized by the letter Z. Microphones are designed either as high-impedance or low-impedance units, though some of them can be switched from high-Z to low-Z. A particular tape recorder requires either one or the other.

While most of the cheap microphones supplied with home tape recorders are omnidirectional (that is, they pick up sound from all directions), many dynamic mikes can be obtained with cardioid characteristics. This means that they pick up sound mainly from the front and are relatively deaf toward the sides and rear. In home recording, this can be a marked advantage because it reduces the blurring effect of sound reflections from nearby walls, and in recording concerts at an auditorium, it reduces unwelcome hall echoes and audience noise.

RECORDING HINTS

Making a good recording, particularly of a live musical performance, is a source of pleasure and pride comparable to taking a good photograph of someone you'd like to remember. And just as in photography, there are a few tricks and techniques to recording which it is well to keep in mind.

The following list of simple Dos and Don'ts is by no means a full compendium of recording technique, but these simple hints can go a long way toward transforming your live takes from mediocre mementos to convincing replicas of musical reality.

DO experiment with different microphone locations until you

find one that best suits the acoustics of the room in which you are recording and the kind of music to be performed. As a general rule, avoid putting microphones close to walls or other sound-reflecting surfaces. This will tend to blur the recording. A free-standing microphone in the central area of a room usually yields better sound.

DON'T put the mikes too close to the performers. A minimum distance of several feet is recommended for most kinds of classical music. Vocalists in particular should be permitted to project their voices at the microphone over a reasonable distance. Crooning into the mike is strictly pop stuff. Violinists and pianists, too, generally benefit from more distant recording which does not overly emphasize bowing scrapes or hammer impact.

DO shut off all noise-makers during recording sessions. Disconnect refrigerators, air conditioners, and fluorescent lights, and turn off steam radiators. Normally you don't notice such regular household distractions, but as an unscored *obbligato* on your tapes they become highly obtrusive.

DON'T put your microphones too far apart when recording in stereo. This creates a very unnatural effect, especially when you are recording a single performer. For example, to record a singer or violinist, place the mikes about six to eight feet apart and let the performer stand midway between the mikes about four to six feet behind the imaginary line connecting the two microphones. An accompanying piano should be the normal distance behind the soloist. If you are recording a group of singers or a string quartet, increase the above distances by about 50 percent.

DO put a rug or a typewriter pad under the microphones. This prevents floor vibrations set up by footsteps or passing traffic from reaching the mikes.

DON'T change the volume level while recording. Adjust the volume level before you start the tape rolling. Ask the performers to play or sing the loudest passage of the piece to be recorded and set the level on the recorder so that the meter swings to the zero mark on the scale (or so that the green lights just come together). Once this adjustment is made, softer passages will naturally fall into their proper place on the relative loudness scale. There is no need to boost them by turning up the volume.

Never, never use your microphone when copying a record onto tape or preserving a broadcast. Use a direct electrical connection from your amplifier to your tape recorder. Otherwise, your recording will be marred by all the distortion in the loudspeaker and the microphone, not to mention the noise and echo within the room where you're recording.

12 Putting It Together— Hooking Up the System

In my bachelor days, I could usually keep my appetite for home cooking satisfied by getting my friends to feed me a nice dinner in return for hooking up their hi-fi systems. It was a very easy way to earn a meal, for there is nothing complicated about hooking up components. It requires no special expertise other than being able to read the instructions invariably furnished with each of the components. In case of doubt or undue timidity, your audio dealer will set up your system for you. But that's going to cost you more than just a bunch of groceries. So, if among your friends you can't find a hungry audio fan, here's how you can easily do the job yourself.

SPEAKER PLACEMENT

Most components can be put wherever it seems convenient. Only speaker placement is critical. In fact, one of the important assets of components or compacts—compared to all-in-one consoles—is that they permit you to locate the speakers at that place in the room where they sound best. It is surprising how many component owners don't take full advantage of this fact. Since optimum placement in a given situation depends on the size, shape, and furnishings of your listening room and the sonic characteristics of your speakers, there can be no hard and fast rules to fit all cases. The major consideration, of course, is getting the most natural and balanced sound possible from your speakers.

You can get more bass, for example, just by moving the speakers into corners. Putting the speakers right down on the floor (the woofer end of the cabinet should be in the lower position) also reinforces bass. The reason is that the wall and floor surfaces adjoining in the corner help project the lower frequencies more efficiently into the room, almost like a horn. You can achieve the same bass-boosting effect by mounting your speakers in ceiling corners, a handy arrangement in small rooms with a shortage of floor space. Wall brackets serve nicely for this purpose.

Bass-reinforcing placement may be a boon to inexpensive systems whose smaller woofers tend to be bass-shy. However, with full-size speakers, such placement may result in an overly heavy, unbalanced bass. In that case, it is best to keep the speakers out of corners and perhaps off the floor.

To operate efficiently in the lower bass range, a loudspeaker should be at least fifteen to seventeen feet from the wall toward which it faces. This allows a sound-projection path long enough to accommodate half the wave lengths of the lowest musical notes and lets these deep tones come through more powerfully. That is why, especially in rooms of moderate size, you often get richer sound by placing the speakers against the short walls so that they face the full length of the room.

Since the better speaker systems available today have good high-frequency dispersion—they radiate the treble frequencies over a broad angle—placement is not super-critical as far as the higher treble frequencies are concerned. However, if your speaker system seems to lose highs as soon as you sit down, you had best set up the speakers so that the tweeters are at ear level.

Room Acoustics

Room acoustics is a second factor to be considered when placing speakers. In addition to their effects on the overall balance of highs and lows, these resonances color to a greater or lesser degree all the sound produced by your speakers. Every room has resonant peaks and dips at certain frequencies and locations, depending on the room's dimensions, giving it what might be called tonal personality. The tonal character of a room is determined by its size, its shape, and its sound-reflective qualities. These factors decide the duration of an echo in the room, which acoustic engineers call reverberation time, and the pitch of the notes that are predominant in the echo. A long reverberation time makes for a spacious sound, and the more reflective surfaces there are, the more the higher tones are emphasized.

Hard surfaces, such as plaster walls, tile floors, and windows, reflect sound the way a mirror reflects light and make a room acoustically "live." The proper amount of reflection gives the sound a pleasing brilliance and richness. Too much reflection causes shrillness and jumbles the music unpleasantly by prolonging each note with excessive echo. Moreover, the music seems to come from all sides at once. The other extreme, too little reflection, makes the sound lackluster and dead.

As a rule, you cannot alter the size and shape of your listening room and thus cannot control its reverberation time. Fortunately, the reverberation effect of the concert hall is—or should be—contained on your records, so that the impression of spaciousness can be obtained even when you play records in rooms of moderate size. What you can do, though, is experiment with the balance of sound reflection and absorption in your room. If the music seems overbright and has a harsh, ringing quality, you need something to soak up some of the sound. Put a hanging or a wall rug over the wall that faces the speakers, or put up some heavy curtains. Anything that is soft will help: pillows, overstuffed furniture, rugs, and the like. These cut down the amount of sonic energy bouncing about the room and suppress excess high-frequency tones.

If the music seems stifled, lacking tinkle in the highs, try pulling back a rug to expose more floor area. Or you can take down some draperies or put up a large, glass-covered picture. Such simple measures can accomplish remarkable results.

In some rooms, problems of resonance make certain notes (usually in the bass range) sound louder than the rest of the music. This is caused by so-called standing waves that make the room act as a resonator. Sometimes it is possible to prevent the formation of standing waves by angling the speakers so that they do not project sound parallel with or at right angles to the walls. Try also to place your speakers where they will be least likely to provoke undesired resonances, or arrange your preferred listening chair so that it is not in a resonant area. This may mean shifting the speakers and the chair along the long walls of a room (or moving them away from the walls) until you find the location at which there are the fewest problems.

The last factor to consider is stereo separation. The speakers should be at least eight feet apart. You'll get the best stereo spread if your listening chair is about equidistant from both speakers. But you can still get a fine stereo effect even if you're sitting off-center. In fact, with good speakers, you'll notice the stereo spaciousness almost anywhere in the room, and therefore you don't have to be hidebound about this kind of musical geometry. Besides, the amplifier's balance control permits adjustment of the relative speaker outputs if the layout of your room makes radically off-center listening more convenient.

Listening Location

In searching for the optimum stereo effect, the general rule is to separate the two speakers so that they subtend an angle of roughly thirty to forty degrees as seen from the listening position. But since each room has its individual characteristics and since furniture arrangements are rarely alike, this general rule is subject to all sorts of variations. More often than not it serves merely as a starting point for experiments. For instance, you can put the speakers farther apart and compensate for the added separation by means of the blend control provided on some amplifiers. Or, if your room is so narrow that you cannot separate the speakers far enough, you can place them at right angles against two adjoining walls with the listening area approximately at the intersection of the two sound-projection lines.

It is also possible by means of speaker placement to emphasize either the directionality or the depth of sound that together make up the stereo effect. Directionality is stressed if the speakers face directly toward the listener. But if your preference runs toward greater depth of sound, with music seeming to fill the whole room without a discernible source, try omnidirectional speakers, or angle your conventional speakers outward toward the nearest wall so that they face away from each other and their sound reaches the listener only on the rebound. This method is especially effective in enhancing the sense of acoustic spaciousness in small rooms, although some stereo separation may be lost. In high-ceilinged rooms or under a gabled roof you can even turn your speakers on their backs so that they face upward and their sound is reflected from above. Though this also reduces stereo separation, the use of reflected sound widens the area of the stereo effect so that the location of the listener becomes less critical.

Not that the listener position for stereo is as critical as many people think. In stereo's early days, it was a commonly accepted half-truth that the listener had to sit at equal distance from the two speakers to hear the maximum stereo effect. Such a listening position is comparable to a center-aisle seat in the concert hall, and it puts you in the location with the most balanced right-and-

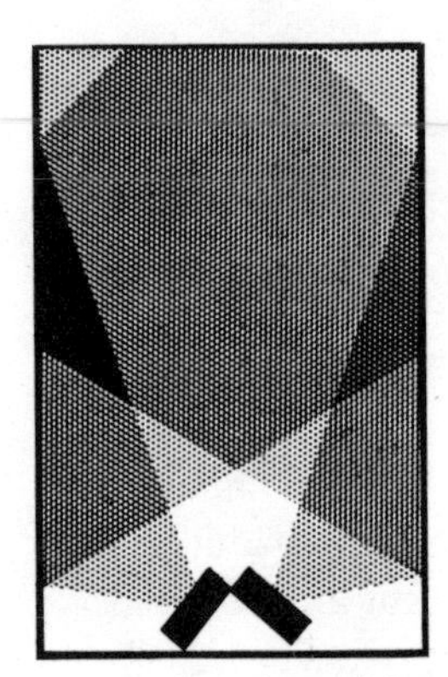

(A) Speakers placed so that sound reflects from the walls of the room.

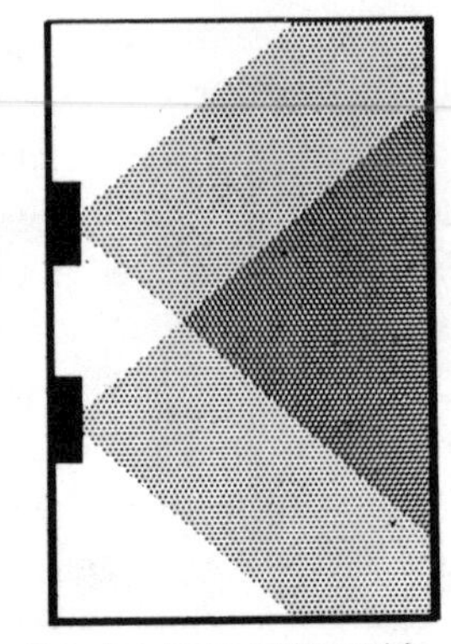

(B) Speakers placed to project toward the short dimension of the room.

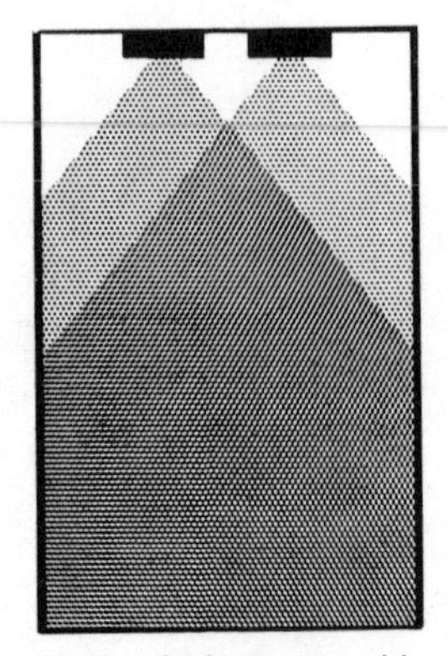

(C) Speakers placed to project toward the long dimension of the room.

Three typical stereo speaker placement patterns, schematically shown as seen from above. Overlapping areas of sound projection are best for stereo listening.

left sound distribution. In a room of average size, the stereo speakers might be placed from eight to twelve feet apart, and you could then put your favorite chair across the room from the loudspeakers at a point somewhere close to an imaginary line drawn midway between them.

But you needn't be dogmatic about your stereo seating arrangement. Stereo permits the listener far greater freedom of movement than orthodox stereophiles will admit. In fact, the stereo effect can be appreciated almost anywhere in the room. If you are sitting off-center, your location might be compared with a box seat along the side of an auditorium, which is nothing to complain about. You may perceive less left-right directionality in such a location, but the essential fullness and spaciousness of stereo will still be retained.

To prove to yourself that the stereo effect is not strictly localized, just walk across the room in front of the speakers while a record is playing. It's like dancing across a ballroom in front of the bandstand: although the sound perspective changes at various points, depending on which instruments you are closest to, these are quite natural changes and the stereo effect is not lost.

Spreading stereo sound evenly over a wide area requires a wide angle of treble dispersion from your speakers. Loudspeakers differ considerably in their ability to fan out the high frequencies. The poorer ones project the treble in a narrow beam, as from a flashlight. This leaves at the sides of the beam large areas of aural "shadow" in which the sound is dull. In general, broad-angle dispersion of highs provides greater latitude in the choice of listening location.

Four-Channel Placement

Although the discussion of speaker placement applies just as much to four-channel as to stereo installations, quadraphony does add some factors of its own. The main problem is arranging your listening position so that the rear speakers are balanced in respect to the front speakers. Moving your listening chairs to the center of the room would be the ideal solution. If your living room is big enough, I recommend it. But for most of us, with average-sized rooms, a clump of furniture in the room's center would entail insurmountable traffic problems, not to mention aesthetic atrocities.

Practical room arrangements usually seat at least some listeners along one wall, and we must allow for this. I've gotten good results by placing the rear speakers right along the couch on which I listen, with the front speakers facing me across the room. But instead of facing the rear-channel speakers straight forward, I angle them outward toward the side walls (away from the couch). Then I adjust the volume so that the sound from the rear speakers does not overbalance the front speakers.

PLACING COMPONENTS

Once the positions of the speakers and your listening chair have been established, you are ready to locate the rest of your equipment. It should be as close to your favorite listening chair as possible, so you don't have to stride clear across the room every time you want to tune in another station or change the volume.

Ventilation is seldom a problem. About an inch of clearance at the rear will properly cool tuners, amplifiers, and small recorders. Very powerful amplifiers or receivers may require a bit more clear space for cooling ventilation. If so, their instruction manuals will spell out their requirements.

In general, your components should be reasonably close together, and the cables between them should be no longer than necessary. "Necessary," in this case, means long enough to reach from one component to the other, plus enough extra at each end to let you pull the component away from the wall to check or change its rear-panel connections. Too long a cable may pick up hum and noise, or diminish high-frequency response a bit.

INSTALLING A TURNTABLE

Turntables should rest within their own spring-mounted base on a steady surface, to prevent vibration from shaking the stylus in the groove. A heavy piece of furniture, such as a sideboard, whose legs sit firmly on the floor, makes a good turntable support. A heavily-laden wall shelf makes an even better one. Make sure that your turntable won't be disturbed by vibrations from the speaker, or by your footsteps as you cross the room.

A turntable with a magnetic cartridge should be connected to the "Phono" inputs of your amplifier or receiver. Its AC power cord may be plugged into an unswitched AC convenience outlet on the amplifier's or receiver's rear panel, but never to a switched outlet. Otherwise, turning off the amplifier while the record player is still on may keep the player from disengaging its rubber drive wheels before stopping. This may cause flat-spots on drive wheels, giving rise eventually to thumping and other noises.

Most turntables also have an extra wire (usually green) for grounding to the amplifier. Most amplifiers have connections marked "ground" for just this purpose. If your amplifier or receiver doesn't have a ground connection for this wire from the record player, you can simply attach it to just about any screw on the chassis. In most cases, such a connection will cancel the hum that might otherwise be caused by the turntable motor.

Tuners and tape recorders are to be connected to the amplifier or receiver at the terminals correspondingly marked. Suitable connecting cables are usually provided by the manufacturer of the components. If not, you can buy these cables—complete with preassembled plugs—at any audio dealer's.

CONNECTING THE SPEAKERS

Most components are interconnected by plug-in cables. But speakers are connected by ordinary lamp cord, usually fastened to screw terminals.

Choosing the right wire is important—too thin a wire can reduce the bass response or overall volume of your system. Regular eighteen-gauge lamp cord, available at any hardware store, will do for lengths of up to ninety feet or so for 8-ohm speakers. Sixteen-gauge wire (which is thicker than eighteen-gauge) should be used for lengths up to about one hundred and fifty feet, and you can use the next smaller wire size, twenty-gauge, for lengths up to sixty feet. For 4-ohm speakers, the maximum permissible length for wire of a given gauge is *half* what it was for 8-ohm speakers, but for 16-ohm speakers you can use double the wire length.

Make all your wire connections carefully—a stray strand of wire causing short-circuits between terminals can damage your amplifier. This is the one operation in the whole set-up procedure where you have to be neat and careful.

It's also important that both speakers operate "in phase." This means that they must work together in tandem, their cones pushing and pulling at the same time rather than working against one another. Conversely, when one speaker pushes forward while the other pulls back, the speakers are said to be out of phase, which usually causes loss of bass and an uneven sound spread.

The easiest way to check your speakers for phasing is to play music with the bass control turned all the way up, then reverse

The only critical step in hooking up a sound system is connecting the speakers to the rear of the amplifier. Be careful that stray wires do not touch the adjacent screw terminal, as this might cause a damaging short-circuit.

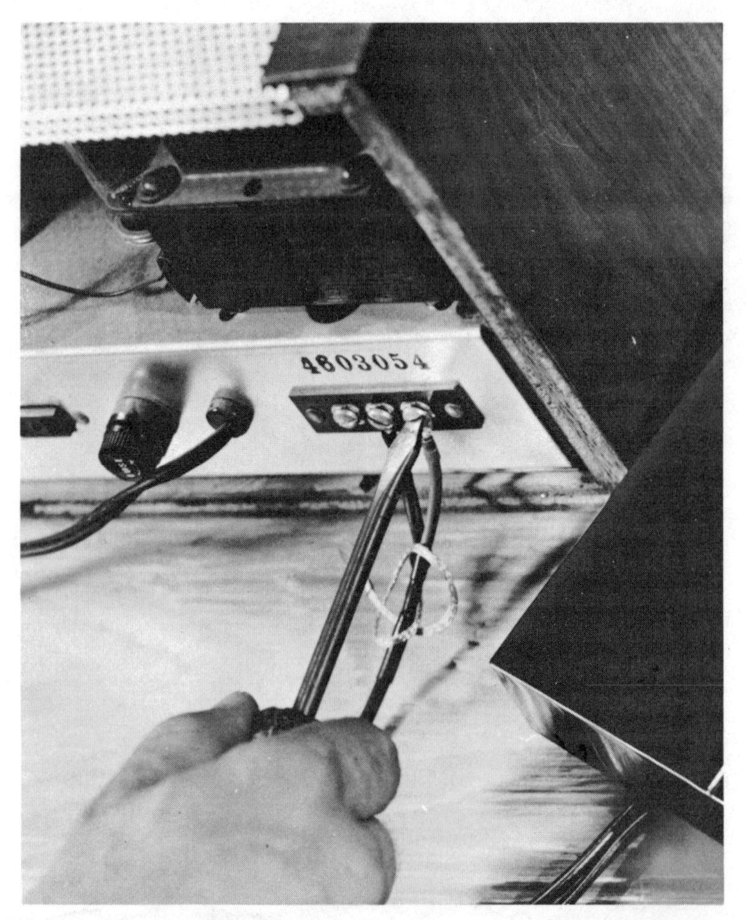

the leads to one speaker. If the bass seems to increase, then you originally had the system misphased; if it decreases, your original connections were correct, and you should restore them. If you can't hear any difference, try moving one speaker till it's adjacent to the other, and repeat the test.

If you're using ordinary lamp cord as speaker wire, close examination will reveal a raised "bead" along one edge of the wire. You can use this as a guide to phasing. If, after wiring one speaker, you see that the "beaded" wire is connected to the "C" or "G" amplifier terminal at one end and the left-hand terminal of the speaker at the other, then you can ensure correct phase by wiring the other channel in exactly the same way. After connecting the "C" or "G" terminal for both speakers, connect the other wire in each pair to the terminal with the number (4, 8, or 16) that corresponds to the impedance rating of the speaker.

ANTENNA CONNECTION

Every tuner or receiver has some sort of built-in antenna that will permit you to pick up nearby stations. But picking them up clearly is another story, especially in stereo. People are all too ready to blame their tuner, or perhaps the station, for fuzzy sound in stereo broadcasts, when in fact it is usually the fault of their antenna. No tuner can perform at its best without a proper antenna, and a mere length of wire dangling from one of the antenna connections is not a proper antenna.

Time and again I have noticed how people cheat themselves of full enjoyment of music over the air simply because they never consider the adequacy of their FM antennas. I have already touched on this matter in the chapter on tuners, but at this point, dealing with the final installation of the system, some elaboration is necessary.

Especially the quality of stereo reception gains a great deal from a good antenna, for stereo requires a stronger signal than mono does to achieve adequate noise suppression, and only in the strongest signal areas can simple indoor antennas provide acceptable stereo reception.

Another disturbing factor, multipath distortion, assumes added importance in stereo. Multipath reflections of a TV signal result in double outlines and blurred images. In FM reception multipath produces a rasping quality in the sound and a loss of stereo separation. Multipath distortion is caused by nearby buildings, metal structures, or hillsides which reflect FM signals in a series of echoes. These echoes hit the antenna fractions of a second later than the direct signal and in erratic phase relationships with each other and with the direct signal. The result is a mishmash of mutually interfering signals that make hash of the musical waveform. Multipath (which occasionally affects mono signals also) can make stereo signals totally unlistenable, because good stereo reception requires the maintenance of accurate phase relationships among the parts of the very complex stereo broadcast signal.

One way to foil this kind of sonic mischief is to install a directional antenna. As the name implies, antennas of this type are sensitive only to signals arriving from one particular direction. When such an antenna is aimed at the station you want to receive, it picks out the signal arriving directly from the transmitter and rejects flanking echoes. If all the stations you normally listen to lie in one direction from your house—in a nearby city, for example—you simply point a directional antenna that way and leave it. But if you want to tune in stations from various directions, you may have to install an antenna rotator, which makes it possible to orient the antenna toward the different stations you want to hear. Directional antennas usually look like the kind of TV antennas you see in outlying areas, consisting of sets of parallel metal rods. A good antenna may have anywhere from three to eight such elements.

The outward resemblance between TV and FM antennas leads many people to believe that they can be used interchangeably. Not so. Although there are a number of antennas designed to cover both TV and FM frequencies, don't assume that an ordinary TV antenna, no matter how complex, will do a good job on FM. If you live in a strong signal area and are suffering from multipath effects, it may be sufficient to replace the fixed flat-line folded dipole antenna that came with your set with a rabbit-ears type of TV antenna. If each leg of the antenna is extended about thirty inches, you will be tuned to the middle of the FM band and you can try adjusting both the direction and the angles of the legs for the best compromise between maximum pickup and minimum multipath. But an antenna designed specifically to cover the FM band—aside from whether or not it will also cover the TV frequencies—is still your best assurance of getting good FM stereo sound from all stations.

This brief guide to installation must necessarily confine itself to general principles and cannot possibly cover individual details that vary with the specific design of specific components. For these, consult the instructions furnished with each component. This is especially important in adjusting the correct tracking force of the tone arm in your record player. But, in general, the foregoing hints should help you get the various parts of your system linked up in proper playing order.

Having completed your system hook-up, you can at last settle back to enjoy your components. After a few days you will find yourself forgetting all about the hardware, just listening to the music. That, perhaps, is the ultimate test for stereo equipment, that it makes you unaware of its presence, so that you feel there is nothing between you and the music.

13 Headphones and Outdoor Speakers

Two optional items, headphones and outdoor speakers, may greatly enhance your enjoyment of your sound system. Outdoor speakers, of course, are useful only if you are lucky enough to have your home surrounded by an expanse of lawn or garden. But even if you live in an urban cubbyhole—in fact, especially if you live in cramped quarters—headphones offer an attractive option: instant privacy.

HEADSETS

Suppose you have an evening set aside for a date with Beverly Sills. Naturally, you don't want anyone butting in. Least of all the kids. But where can you sneak off to be alone with Beverly? Time was when well-appointed homes sported a separate music room where one could take a vacation from the family behind sturdy oak doors. But in this split-level age of cardboard walls and doorless togetherness, it's getting harder all the time to find that island of quiet necessary for the full enjoyment of music.

A growing number of hi-fiers rely on earphones to remove themselves from the scene—psychologically at least. Slip them on, and you are suddenly whisked away from all petty distractions to your rendezvous with Beverly. Besides, with earphones —or headsets, as they are now often called—you can stretch out your date far into the night. You can still listen at full volume after everyone, including the neighbors, has gone to bed.

Privacy, of course, works both ways. While Sills thrills you with high C's, your wife may be trying to entertain a thought or two of her own. So she, too, appreciates it if you keep Beverly to yourself.

Headset Illusions

Aside from being instruments of domestic peace, headsets have yet another advantage. Because the sound goes directly to your ears, it skips all the acoustic quirks of your living room. Problems like speaker placement and stereo listening location are automatically bypassed. The acoustic image of the place where the original recording was made reaches you without being altered by your own home acoustics. The net effect is an uncanny illusion: the very space of the concert hall or recording studio seems infused via the stereo earphones right into your head.

Donning stereo earphones for the first time nearly always causes a reaction of total amazement. You can "feel" the whole concert stage—maybe some sixty feet wide—spreading out in whatever space there happens to be between your two ears! With the whole orchestra inside your cranium, you'll find it hard to believe that your hat size stayed the same.

How does this happen? Nobody knows for sure. The basic process of sensory perception is still not completely understood. Psychologists, physicists, and philosophers have yet to discover the exact relation between objective reality and our senses. Of course, it may be argued, as did Immanuel Kant, that all perception ultimately lies within our heads. The sensation of hearing stereo "space" via earphones is surely an astounding demonstration of this.

Getting back to the more tangible subject of hardware, the recent comeback of headsets was first greeted with loud snorts: "Earphones? I thought they went out with catwhisker radios!" Today's stereo headsets are a wholly different breed. While their early ancestors were nothing but primitive telephone receivers with a metal diaphragm stretched across a signal-activated magnet, modern earphones—quite aside from being two-channel stereo devices—are crafted with the same precision that marks today's audio components. Structurally they resemble full-sized speakers, consisting essentially of a permanent magnet, a voice coil, and a carefully designed and suitably suspended cone. By way of analogy, one might say that today's stereo hi-fi headsets compare to ordinary radio earphones as a racing yacht compares to a tree-trunk canoe. Mainly the difference lies in calculated refinement.

Examples of this refinement include headphones with separate woofers and tweeters (and easily adjusted tweeter level controls), with stereo-mono selector switches, volume, balance and tone controls; electrostatic headphones (now far more common than electrostatic speakers); and even four-channel headphones.

If you've had no prior experience with modern headsets, you will be amazed at the full bass obtained from such small sound generators. How can low frequencies be so effectively reproduced by diaphragms measuring only about two inches in diameter? Ordinary loudspeakers must be relatively large for adequate bass response because they must push plenty of air to project bass energy into a room-size listening space. But the "listening space" to be filled by earphones is only the tiny air volume between the earphones and your eardrums. Moreover, with the headsets fitted tightly against your ears by means of soft padding, this small air space is sealed off and represents what the engineers call a "closed system." This provides practically loss-free transfer of low-frequency energy. Under such conditions even a small sound generator suffices to create ample bass.

Like loudspeakers, stereo headsets tend to have their individual sound coloration. In selecting a model for your own use, compare different designs just as if you were buying a pair of speakers. Clarity is the most important criterion. Make sure the sound doesn't blur even at full volume. Try to pick out the individual instruments in the orchestra. Watch for the presence of bass even in soft passages; note if the sound of the lower strings has its proper solidity. Check the transient response by listening critically to the crispness of sound in such instruments as harpsichord, guitar, and various kinds of percussion. And watch for that common drawback of inferior designs—high-frequency distortion. Violins, for example, should sound silky and smooth, without stridency.

Headset Fit

Aside from sound, fit is the main factor in picking your earphones. You should be able to wear your headset all evening without any discomfort. Fit around the head is rarely a problem because most headbands are either flexible or otherwise adjustable. The earpieces, however, have fixed dimensions. So make sure they don't pinch or squeeze your ears. They should fit around the ears rather than over the ears. They should also provide a good air seal; otherwise you lose bass response. Some ear cushions are liquid-filled so that they mold themselves to the contours of your head. Others rely on foam materials to form an efficient sound seal. On some models, ear cushions are washable, a decided advantage in case they become grimy or saturated with skin oils.

The lighter the headset, the longer you can wear it without fatigue. For this reason, virtually all recent models are made of lightweight materials. Some weigh as little as 9½ ounces or less. You are hardly conscious of wearing them.

FIDELITY AL FRESCO

Headphones expand your listening horizons internally. But, neighbors permitting, there's a way you can extend your listening into the great outdoors, enjoying music by starlight, symphonies under summer skies. You listen in your lawn chair, your receptiveness enhanced by nature and a tall drink. Or, if you are a social type, picture yourself as the shirtsleeved impresario of an intimate, informal soirée right in your own garden or back yard. You pick the players, the program, and the guests. And during intermission, you're the cook. You might even stage your own summer theater productions, with a repertoire ranging from Broadway musicals to Shakespeare. All it takes is agreeable (or distant) neighbors, and a pair of outdoor speakers hooked to your sound system.

There's no point trying to lug your heavy indoor speakers out into the open. You'd only risk damage to their outsides and your insides. Besides, you're better off with speakers specifically designed for outdoor use. You don't even have to nestle them in a protected spot under the eaves. Leave them right out in the open, sitting on a lawn or hanging in a tree. They are completely weatherproof. Neither rain nor wind, nor owl, nor squirrel can do them damage. In fact, these speakers are so impervious to the rigors of outdoor life that you can leave them out all year. They'll thaw out in spring.

These speakers owe their ruggedness to special materials combined with special design. Typically, they are enclosed in plastic or fiberglass shells doing double duty as shields and acoustic baffles, and the speaker cones and other working parts are moisture-proof.

Outdoor speakers as such are nothing new. You've heard them at railroad stations, in ballparks, and on sound trucks at election time. But you wouldn't want those tinny squawkers rasping against your ears when you're playing music. To fill the growing demand for outdoor speakers with good tone quality, a handful of hi-fi manufacturers have come up with weatherproof models capable of doing fair justice to music.

Placement and Hookup

The rules for setting up outdoor stereo are basically the same as for indoors. To get the optimum stereo effect, the main listening area should be located about midway between the two speakers. But to get the sound spread over the usually larger outdoor area, the distance between speakers must be increased. Mine, for example, are about fifty feet apart. One is up a tree, the other hidden in a lilac bush. Of course, you can let your speakers just sit on the lawn. A friend of mine has his leaning on a couple of garden dwarfs. Most models come equipped with mounting brackets so they can easily be attached to tree branches, under the eaves of your roof, or any other wooden surface.

Many of the newer stereo amplifiers and receivers have a special set of output terminals for extension loudspeakers. In that case, you just connect your outdoor speakers to these terminals. Amplifiers or receivers with extra output terminals usually also have a speaker selector switch to let you choose between indoor and outdoor speakers, or play both at the same time. If your amplifier lacks these features, most hi-fi stores carry switches you can add to your system for about $5–$10. But don't try connecting two pair of speaker lines to the same terminal screws at the rear of the set; connecting all the leads to the same screws may bunch up the wire strands around the terminals. This often causes shorting between neighboring terminal posts.

As an added refinement, you may want to put "L-pads" into your outdoor speaker lines. These act as separate volume controls for your outdoor speakers. You can locate these controls close to the speakers themselves, so you don't have to run indoors to the amplifier every time you want to change the outdoor volume. These L-pads, too, are available from most electronic mail-order firms.

Whether you just string a pair of wires through the window for connecting your outdoor speakers or make a more permanent connection through the wall is mainly a matter of taste and convenience and will not affect performance one way or another. Certainly the neatest way to do it is to lead the wires through the wall, perhaps to a weatherproof connector on the outside of the wall where your outdoor speakers can then be plugged in. The outdoor wire, whether you bury it or leave it lying in the grass, should be the kind with weatherproof insulation.

If your amplifier does not have built-in provisions for hooking up an extra pair of speakers, you must be careful about one thing. Some amplifiers may be overloaded and damaged by having too many speakers connected to them, particularly if the speakers happen to be of unusually low impedance—say, 4 ohms instead of the customary 8. If your particular amplifier is susceptible to such damage, a suitable caution would be given in the manufacturer's instructions. So be sure you read them carefully before hooking up the extra speakers.

Once your outdoor music system is all hooked-up and working, you will notice a surprising musical phenomenon: records sound better outdoors than live music does. Live outdoor music usually suffers from the absence of the reflective surfaces which reinforce music heard indoors; it sounds thin and dry. Records, by contrast, carry in their grooves the indoor acoustics of the concert halls or studios where they were made. It sounds as though the walls of the concert hall were standing invisibly in your garden.

14
Life Insurance for Your Records—a Matter of Love *and* Money

When you buy a record it usually means you like the music so much that you want to have it for keeps. But how long will it actually last? And how good will it sound after the first few plays?

Preserving your records takes on a particular urgency when you consider that almost any record collection is irreplaceable. Most records go out of print so quickly that often you cannot buy another copy of a favorite disc after it wears out. Any effort to prolong the life of your record also makes good economic sense. In aggregate, your record collection probably constitutes a considerable investment, greater by far than your investment in components. Wearing out your treasured records therefore is a loss of both love and money. Finally, it is self-defeating and irrational to set up a superior sound system only to spoil the effect by playing records full of irritating blemishes; for the better your system, the more faithfully it will render every scratch and every speck of dirt on your discs.

KEEPING YOUR RECORDS CLEAN

Of all record killers, plain household dust is the most vicious. That is why I regularly include the following pious motto among my New Year's resolutions: "From this day forward I shall clean every record before each playing." Nothing else, I assure you, will as effectively lengthen the life and preserve the pristine sound of your discs.

A coddled LP, cleaned before every playing, sounds almost as lush at the hundredth playing as at the first. But a grimy, neglected disc may scrape itself out in less than twenty plays, the texture of its sound turning from silk to sandpaper, its pianissimos struggling against a rising level of background noise, and its musical phrases senselessly syncopated with clicks and crackles. Components that respond to the subtlest nuance of timbre in the record groove will just as faithfully render every snap, crackle, and pop produced by seemingly negligible amounts of household dust in the grooves.

I realize that the notion of a mighty symphony foundering on a few specks of dust seems most unlikely. But look at the situation from the music's point of view—down in the bottom of the record groove, a zigzagging valley whose narrow twists and turns are the physical equivalent of the musical sound waves. The tip of your stereo stylus races along this crooked path with astonishing force. The downward force of the stylus may be only a small fraction of an ounce, but the weight is concentrated on the two exceedingly tiny contact areas of the stylus tip. The effective pressure of the stylus in the groove is, therefore, equivalent to thousands of pounds per square inch. Suddenly a dust particle looms in the path of the stylus like a large hard rock with razor-sharp edges. The stylus crashes against it, and something evidently has to give. That something is the soft vinyl groove that holds the mighty symphony.

The sound of this collision may be but a tiny click in your speakers, but thousandfold repetition by thousands of dust particles spreads a sonic haze over a once-brilliant recording. The tonal gloss that the recording engineer so successfully transferred to the grooves gives way to ear-grating shrillness. Dirt, by the way, puts a double hex on stereo. If the stylus is thrown off by dust particles, it responds in two ways to the detour: with a lateral and a vertical signal. In effect, you are listening to the noise in stereo also, and this inevitably results in a far greater sonic irritation than the old-fashioned mono noise.

Most amplifiers, to be sure, have a scratch filter designed to remove these unwanted sounds when noisy records are played. But this is hardly a satisfactory answer to the problem, for along with the dust-caused surface noise, the filters also clip off quite a bit of the high-frequency range that is the hallmark of lifelike sound. Therefore, at least in phonographic matters, cleanliness is next to godliness.

But virtue seldom comes easy. Keeping records free of dust is as hard as keeping a blue serge suit free of lint, and for exactly the same reason: static electricity. Dust clings to records with the passion of a determined lover, and it is almost impossible to brush it off. In fact, brushing may only strengthen the close relationship between disc and dust by increasing the static charge. The record has to be tricked out of this sonically fatal misalliance. One way to do this is to coat the record with a thin conductive film that will neutralize the static charge. A number of such de-staticizers are available. Some are applied with a velvet pad, which simultaneously cleans the record; others are simply sprayed from aerosol cans. The problem with most of these preparations is that they leave a residue in the grooves. The heavier-tracking pickups of a few years ago had no trouble plowing through the gum-like residue and making contact with the groove wall. But modern lightweight cartridges with highly flexible styli—tracking at 2 grams or less—are often derailed by the gunky remnants of antistatic fluids.

To avoid such complications, I prefer to use plain water as a de-staticizer. Applied sparingly, it provides a film of moisture which dispels the static charge when I clean the disc and later evaporates entirely. I have found an inexpensive cleaning gadget called Record Preener (sold by Elpa Marketing Industries for $4.00) to be very effective. It has a velvet cleaning surface, which is dampened by an internal moisture wick. The nap of its velvet apparently has just the right length and resilience to bring up the dust all the way from the bottom of the groove. (This is important, for a light surface cleaning of the "land" between the grooves—an area that the stylus never touches—does nothing to remove the noise-producing substances in the grooves themselves.)

HANDLING YOUR RECORDS

Next to cleaning the disc before each play, careful handling is the most important factor in record care. This means a hands-off policy. You may not think of your hands as greasy paws, but the fact remains that your fingers deposit an oily film wherever you touch a record. This film, in turn, gathers airborne dust and turns it into grime. Keeping your fingers off the grooved part of the record is the best method of preventing the formation of this grime. To hold a record in one hand, support it beneath the label with your index, middle, and fourth fingers while keeping it steady by pressing your thumb against the record rim. With a little practice, such "sanitary" record handling becomes habitual. It might even be a good idea to hold a few training sessions for every member of your household who has access to your record collection.

CLEANING RECORDS

The record hygiene recommended here is mainly preventive. To restore a record already scarred by neglect is rarely possible after the dust has done its damage. But, as radical therapy, you might try washing the disc with cool tap water (to which you have added a few drops of liquid detergent). Scrub the record with a sponge, rinse with clear water, and then allow it to drip dry. This should loosen and remove some of the encrusted dirt and thus help—literally—to clean up the sound.

CARING FOR THE STYLUS

Not only your records, but the stylus, too, should be included in your anti-dust campaign. Dirt—plain or otherwise—comes as naturally to a stereo stylus as to a pig: it just digs it up. During the play of a single twelve-inch side, the stylus literally sweeps up about two and a half miles of groove—the curviest, nookiest dust-catcher you ever saw. Dirt mounts in miniature heaps on the stylus tip and tends to derail the stylus on its travel along the record track. In fact, a dirty stylus is the most frequent cause of distortion and groove skipping. Audio repairmen tell me that on more than half of all the service calls they answer, stylus cleaning is the only problem. Obviously, this is something you can and should do yourself.

The best way to clean the stylus is with a special stylus brush, available at nominal cost in almost any specialized audio shop. Some turntables and changers provide a stylus brush for this purpose. Never try to clean the stylus by dragging your finger across the stylus tip. In the new high-compliance cartridges, the stylus is so delicate that a casual touch by your finger would bend it out of shape. This is a good place to warn against the fatal myth of the "permanent needle." There is no such thing, and misplaced faith in the permanence of your stylus may take a heavy toll of your discs.

A diamond is, of course, a record's best friend. In fact, it is the only kind of stylus ever used with the better audio equipment, and this is understandable if you consider that the sapphire styli still found in lesser breeds of phonographs last only about forty playing hours before they begin to chew up records. But even a diamond, contrary to the pronouncements of the jewelry trade, is not forever. Granted, in modern high-compliance cartridges tracking at less than 2 grams, a diamond in average use may last up to five years or even longer. But what happens after that—or possibly sooner? Instead of presenting a gently rounded surface to the groove wall, a stylus that is worn has sharp edges that tend to abrade the delicate contours of the groove. This not only wipes out the high frequencies, but also creates jagged areas that produce noise in subsequent plays.

Similar damage is done by a diamond that has been chipped, perhaps as a result of someone's accidentally dropping the tone arm on the turntable base. Lately, special cuing devices and viscous-damped arms have been incorporated in some record-playing equipment to prevent accidents of this sort. But even so, just to be on the safe side, it's a good idea to take your cartridge to your audio dealer for a close examination under the microscope about twice a year. Wear or damage usually shows up in the form of flat facets on the diamond's conical tip. The dealer can also check your cartridge for mechanical misalignment of the stylus assembly, another frequent cause of poor sound and untimely record wear.

You don't need to lug your whole turntable to the shop to have your cartridge checked. The part of the tone arm which holds the cartridge is detachable. In most models, all you have to do is loosen a threaded retainer ring and the cartridge, along with its mounting shell, pulls right off with a slight sideways twist. But be careful as you do this—and also in taking the cartridge to the dealer—that the stylus doesn't get bent.

CHECKING STYLUS PRESSURE

You should also check the tracking force of your tone arm every few months with a stylus-pressure gauge, an accessory obtainable in audio shops at nominal cost. Most audiophiles are well aware that excessive tracking force hastens record and stylus wear, but it may come as a surprise to many that unduly *light* pressure also causes distortion and may damage records. The stylus-force adjustment should always be set within the range specified by the cartridge manufacturer.

THE RIGHT WAY TO STORE RECORDS

Proper storage is also important in protecting your records. Keep them away from radiators in winter and out of direct sunlight in summer. Don't stack them on top of each other, for the weight of the stack itself can damage the discs on the bottom of the pile. Shelving records vertically (standing on edge and evenly supported from both sides) helps to prevent them from warping.

After playing your records, don't leave them strewn about "naked" on dusty shelves, sofa pillows, or other such convenient parking places. And when you put a record back into its jacket, always slip it first into its protective envelope to keep the cardboard from scraping against the record's sides. Given such care, your records could (according to a report by the Library of Congress) outlast you by more than a hundred years.

15
An Epilogue on Music

All sound equipment is but a means to an end. In itself it is a mute and meaningless assemblage of metal, wood, and plastic. Its meaning emerges only with the realization of music. It is precisely in this sense that audio equipment becomes the key to a profound personal adventure: the discovery and exploration of one of the profoundest forms of human expression.

"Where words fail, music begins," said Heinrich Heine, the poet whose works so often served as inspiration for Schubert and Schumann. And even the dour Thomas Carlyle was not impervious to the enchantments of music: "See deep enough and you see musicality," he wrote in his *Sartor Resartus*.

To write about music is hopeless just because it is an art beyond language. Perhaps the German novelist Hans Henny Jahnn comes closest to paraphrasing in words the essence of music. "Music has no analogies," says Jahnn. "She rejects all comparisons, her colors are without names; there are no loins to still her passions and desires, no death precedes her mourning, and in her palace chambers human flesh and blood dissolve into delight and melancholy."

These thoughts come to mind as I cast about for ideas on how to advise you about building your record library. On reflection, it seems to me that the best I can do is to suggest how *not* to build it. Don't try to build a "balanced" library with some representative samplings from all major composers and every musical style period. This approach may be appropriate for a public library or some other institutional collection. But you don't really have to document the art of music in its historical development. Your only object is to please your own taste. So build yourself an *un*balanced record collection, skewed strictly to your pleasure.

The great variety of current popular music and the inexhaustible treasures of folk song can be a tremendously rewarding adventure of exploration. The classical literature can be approached in the same casual, open-ended way. But since classical music is a more formal body of work extending over many different style periods, the question is where to start. The first step is to find out your musical prejudices—and pamper them. Tune your FM tuner to a classical music station. Sooner or later, something is bound to hit your fancy. Then go to the record shop and explore other works by the same composer. Later branch out to some of his contemporaries.

Are you turned on by Mozart concertos? Then try some of his symphonies, or maybe switch to Haydn, who wrote in a similar style but often with a quite different emotional undertone.

Suppose you've steeped yourself in one style period to the point where you sense all its subtler musical meanings. Then pull a complete switch on yourself. If you started out with Mozart or Haydn, leave the elegant formalism of the eighteenth century for a super-romantic binge among the tone poems of

Multiple microphones hovering above the orchestra convey every nuance of an intricate symphonic score. (*Photo: RCA*)

Richard Strauss. Sample the iridescence of Debussy, the splendid architecture of Brahms, the exuberant austerities of Bach, the American colloquialisms of Copland, or the lean angularities of late Stravinsky. Once you have branched out in different directions, you'll find roads leading to everywhere in the realm of music. And the rewards are varied and infinite.

To cut the cost of your exploratory trips, you might for a start stick with records on the so-called discount labels. These records sell for less than half the price of standard records and are found under such labels as Nonesuch, Seraphim, Odyssey, Turnabout, Victrola, and London's Treasury. These series are comparable to paperback reprints in the book business. They offer excellent material at minimum cost. Yet not all the recordings on these low-priced labels—which sell in some stores for less than $2 per disk—are older vintages. Many of them are brand new, outstanding performances, usually made by European artists who don't happen to be top box office in the United

States at the moment. Consequently, their work does not appear on premium-priced discs, regardless of the artists' merit. Such are the economics of music merchandising. Draw your profit from it.

Find a record dealer who really knows music. Some of them still do. Then take his advice. A really devoted record dealer can sense your own musical taste and match it with appropriate records from his stock. You won't get this kind of service if you buy your records at the local supermarket. But just as there are still a few bookshops left where the owner really cares about books, you can sometimes find a small specialized record store run by someone with a sense of personal commitment to his merchandise and a sense of responsibility toward his customers.

One way to recognize such a shop is by a characteristic air of leisure. Neither customers nor salespeople seem in any hurry. They always seem on the verge or in the middle of lengthy discussions about the relative merits of different recordings, the changing styles of musical performance, or the exploration of newly recorded repertoire. Finding such a record shop—and you can still find them in some of the larger cities—is like joining a congenial club. It becomes a refuge from your problems, a forum for your opinions, and a place for making friends.

However, if you live far from such enclaves of civilization, you can get reliable guidance for your record collecting from such publications as *Stereo Review* or *High Fidelity,* monthly magazines largely devoted to record reviews.

According to popular myth—as persistent as it is pernicious—serious music is inherently forbidding. Yet as your record collection grows, you will find that the *only* requirement for enjoying great music is listening. You don't need to "know about it." To say that music reveals its meaning only to the initiated is like saying that you can't go for a drive without understanding the theory of combustion engines. In every other context, such nonsense would be laughed at. But when it comes to music, lots of people believe it. Think of it this way: If someone tells a story, you can understand it even without knowing the rules of grammar. And so it is with music. Just listen, and you will get the message. After all, the great composers didn't just write for other composers—they wrote for the public.

What then does the music say? And why do we listen? Every composer, of course, speaks with his own distinctive voice. Yet all great music springs from a common source. It takes on meaning by reflecting and transforming our basic emotions. We all carry within us a core of profound and universal feeling. We all, in some form or other, experience love and yearning, regret and grief, striving, struggle, triumph, and defeat. We all are filled on occasion with wonder and a sense of mystery, and all suffer fear of death. These feelings are the common denominator of humanity. They are also the root of music.

In meaningful music our feelings appear intensified, idealized, and heightened. The elements of common experience are tailored and proportioned within the music by the composer's design. One might think that such "processing" of the emotional raw material into a work of art might distort it. Surprisingly, the opposite is true. Music concentrates emotional reality so that we can perceive it in sharp focus. It takes the blinders off our inner vision and jolts us from our workaday rut. Often it reveals unknown regions within ourselves. The music prods beneath our crusted surface, shaking loose undreamed resources of responsiveness and intuition.

For all its pleasure, listening to great music is a rather solid business. It fortifies the soul. In listening, you become part of the music. Some of its beauty, some of its lastingness gets into you. As T. S. Eliot observed: "You are the music while the music lasts." This transformation in itself becomes a creative act on your part—one of those transcendent moments by which each of us marks his existence in the stream of time.

Glossary

A-B tests are comparisons of the performance of two components made by listening first to one and then quickly switching to the other. The sudden contrast clearly reveals any audible differences that may exist between the two units.

Acoustic feedback occurs when sound vibrations from the loudspeaker travel back to the record player. The phono stylus then picks up the vibrations from the speaker along with the modulations in the record groove and feeds both signals to the amplifier and speaker. The result is—at best—a slight tonal blur or a rumbling noise. At worst, the re-amplified vibrations pile up, overload the amplifier, and produce a loud rumbling, humming sound that, if sustained, could damage the speakers.

AFC (Automatic Frequency Control) is used in many FM tuners to lock in a station to keep it from drifting out of tune. Since AFC tends to "pull" the tuning to the stronger of two adjacent stations, an AFC-defeat (on-off) control is necessary to permit tuning in weak stations. Some tuners dispense with AFC altogether, relying on a circuit design that is inherently free of drift problems to hold a station steady on the dial. The absence of AFC in a quality FM tuner is therefore not necessarily a drawback.

AGC (Automatic Gain Control) is a circuit employed in most FM tuners to adjust the amount of amplification to the strength of the incoming signal. This helps keep the tuner from being overloaded by excessively strong signals from nearby stations.

Alignment is the term used to describe certain maintenance adjustments to be made periodically to keep tuners and tape recorders in top working condition. On tuners, alignment involves the adjustment of certain internal circuits; on tape recorders it means adjusting the position of the heads for optimum response.

AM (Amplitude Modulation) is a method of broadcasting in which sound waves are transmitted as variations in the *intensity* of a radio-frequency signal. This method of broadcasting is subject to atmospheric and man-made interference, and in practice is too limited in frequency and dynamic (loudness) range to satisfy high-fidelity requirements. A newer system of radio transmission, FM (or Frequency Modulation), does not have these problems and is therefore preferred for high-fidelity applications.

Amplifier. The amplifier, a basic component in any sound system, receives from the cartridge weak electrical signals representing the music recorded in the record grooves or similarly weak signals from the tuner or tape recorder. It then enlarges (amplifies) these signals to make them powerful enough to drive the speakers.

Amplitude refers to the strength, or loudness, of a sound, or of the strength of an electrical signal representing sound. If the sound or signal is represented as a wave pattern on an oscilloscope or graph, the amplitude corresponds to the height of the wave (i,e., the magnitude of the swing in each cycle).

Baffle is an older term, though still used, for any loudspeaker housing. The main job of the baffle, or enclosure, is to keep the sound waves radiated by the rear of the loudspeaker cone from canceling the sound waves projected from the front. Since such cancellation occurs only at the lower frequencies, an unbaffled speaker invariably sounds bass-shy. For optimum performance, the enclosure must be acoustically matched to the loudspeaker it houses.

Balance refers to the relative loudness of the two speakers used in stereo listening. Balance can be set at several points in a component system (FM-tuner outputs or power-amplifier inputs), but final balance is adjusted by means of the balance control on the amplifier and makes both speakers sound equally loud when heard from the listener's location.

Bass boom refers to an unwanted bass resonance in a loudspeaker system. The effect is like that of speaking into a barrel: the resonance masks the natural tone color of the voice. Similarly, a speaker system whose resonances are too pronounced, or are at the wrong frequencies, falsifies the true character of sound, creating a hollow, thumpy bass that blurs the clarity of the music. Jukeboxes and poorly designed console phonographs are notorious for boomy bass.

Capture ratio measures an FM tuner's ability to sort out two stations operating on the same frequency. A tuner with a poor capture ratio will pick up both stations simultaneously, making it impossible to listen to either. A tuner with a good capture ratio will "capture" the stronger station and reject the weaker one. Capture ratio is expressed numerically; the lower the figure, the better the capture ratio. Don't confuse capture ratio with selectivity. Selectivity refers to the tuner's ability to separate adjacent stations on the dial. Capture ratio refers to the separation of stations on the same spot on the dial.

Cartridge, as the term is usually used, refers to the small device (also called a phono pickup) mounted at the active end of the tone arm. It follows the record groove by means of its stylus (needle) and translates the undulations of the groove into electrical signals. (The term cartridge is also used to describe the plastic-encased tape reels designed for cartridge loading of certain kinds of tape recorders or tape players.)

Channel is the term used to indicate one of the two separate signal and sound paths employed in stereo, hence the expressions "left channel" and "right channel."

Compliance describes the amount of force that must be applied to the stylus of a phono cartridge to deflect the stylus a given distance. Compliance is expressed numerically—for example, 15×10^{-6} cm/dyne. This means that if 1 dyne (a basic unit of force) is applied to the stylus, it will be deflected one 15-millionth of a centimeter. When comparing cartridge specifications, make the first number—the one before the multiplication sign—your basis for comparison. The higher the number, the greater the compliance. Higher-compliance cartridges can be used only in precision tone arms.

Components are the basic units of a sound system: turntable, tuner, amplifier, and speakers. The term is sometimes used to distinguish high-fidelity from low-grade sound equipment.

Crossover networks are used in loudspeakers to separate the treble from the bass in order to feed high-frequency tones to the tweeter and low-frequency tones to the woofer. The simplest type of crossover is a high-pass filter, which merely keeps the bass notes out of the tweeter, thus preventing tweeter damage. A more complex crossover, consisting of one or more coils and capacitors, also keeps the highs from entering the woofer, where they might be distorted. The frequency above and below which the frequencies are routed to the woofer and the tweeter is called the crossover point. A well-designed crossover network must be matched to the characteristics of the speakers so that a smooth transition will occur at the crossover point. For speaker systems with a separate mid-range speaker, a more elaborate three-way crossover network is used to divide the total frequency range into three portions—bass, mid-range, and treble—each going to its appropriate speaker.

Crosstalk is a term originated by telephone technicians to describe interference between two phone calls on adjacent wires. When applied to audio, crosstalk means that some left-channel signal is leaking into the right channel, or vice versa, thereby reducing stereo separation. Commingling of the left and right signals is not wholly avoidable, and is practically unnoticeable if the intruding signal is at least 25 db lower in volume than the signal rightfully belonging to the channel. Crosstalk is most often stated in terms of stereo separation. A separation of at least —25 db (the higher the negative figure, the better) in the specifications of a cartridge, amplifier, or tuner therefore signifies that there will be no audible crosstalk between the channels.

Damping, as applied to a loudspeaker, describes its ability to come to a complete stop the instant the electrical signal that is being fed into it ceases. In a system with poor damping, the speaker cone will continue to vibrate for a moment after the input signal has ended. This "hangover" blurs musical details. Amplifiers also have a "damping factor," which helps control the speaker. A high amplifier damping factor, usually above 10, is satisfactory for most speaker systems, although some speakers operate best with amplifiers that have lower or higher damping factors. Clarity in the reproduction of complex orchestral passages—particularly those involving heavy percussion—

is an indication of good damping characteristics, since good damping contributes to a speaker's transient response.

Decibel, abbreviated "db," is the standard measure of loudness. It is a *relative* measurement, used to compare two different loudness levels. For instance, if a loudspeaker is "5 db down at 40 cycles" (usually written "—5 db"), the sound it produces at a frequency of 40 cycles per second is 5 db softer than the sound it produces at a standard reference frequency—usually 1000 Hz. The smallest readily apparent loudness difference in music is 3 db, though sharp-eared listeners may discern differences as small as 1 db.

Distortion, in playback, is any change in the recorded sound that takes place in the playback system. At its worst, distortion can make a violin sound like a trolley car screeching around a curve. More frequently, however, distortion is quite subtle and barely perceptible at first. But in prolonged and attentive listening it causes a sense of discomfort known as listener fatigue. In recent years, improvements in audio design have reduced distortion in quality components to such a low level that it does not obtrude upon the listener even after many hours of concentrated listening. Distortion exists in two principal forms: harmonic distortion, which falsifies tonal nuances, and intermodulation distortion (usually called IM), which results from the interaction of various frequencies within the playback components and produces a harsh, grainy sound. Precise numerical statement in the manufacturer's specifications of both of these types of distortion is a hallmark of trustworthy sound equipment. In high-fidelity components, distortion is usually expressed as a percentage of the total sound at a certain level of output power. High-quality amplifiers should have less than 2 percent IM and harmonic distortion when operating at full-rated output.

Dynamic range refers to the loudness spread (in decibels) between the softest and the loudest parts in a piece of music. Also, the spread between the softest and loudest sounds a system can properly reproduce.

Efficiency is a term usually applied to loudspeakers and is used to indicate how much amplifier power must be fed into a speaker to produce a certain loudness level. To say that a speaker is inefficient does not reflect on its quality. It merely indicates that the speaker requires a fairly large amount of amplifier power to produce a given loudness. Some speakers (particularly certain bookshelf models) are notably inefficient and need at least 20 watts per channel. As long as sufficient amplifier power is available, this is no problem. However, if you own a low-power amplifier—say, about 12 or 15 watts per channel—make sure, when choosing your loudspeakers, that the amplifier will be able to drive them to adequate listening-room volume without distortion.

Equalization, in a general sense, refers to any deliberately introduced change in frequency response. It is used, for example, during disc recording to boost the treble range and weaken the bass range. During playback, an opposite equalization is applied to restore the original tonal balance. The treble range is emphasized during recording so that when the treble is reduced in playback, the record's surface noise will also be reduced. The bass range is reduced during recording in order to prevent the cutting stylus from overcutting the groove at low frequencies. As the term "equalization" implies, everything comes out "equal" in the end, hopefully with flat overall response between microphone and loudspeaker. Since 1955 all records have been cut for playback with the RIAA (Recording Industry Association of America) equalization curve. Either the playback amplifiers provide equalization in their preamplifier stages, or the cartridge itself compensates for the RIAA curve.

Tape is equalized both in recording and in playback to compensate for the inherent high- and low-frequency losses in the recording process. In addition, there are different equalizations for each tape speed; on most tape recorders, the correct equalization for each speed is automatically switched in when the speed is set.

Feedback is a term indicating that a signal is being returned to some earlier point in the amplification chain. In some cases, the return of the signal is intentional and has salutary results. In negative feedback, for instance, part of an amplifier's output signal is applied to an earlier stage in negative (i.e., opposite) phase to the signal normally at that point in the circuit. Negative-feedback circuits reduce distortion, extend frequency response, and are essential to the electronics of high fidelity. Unwanted feedback, however, can be quite pesky. (See Acoustic feedback.)

Flat response. See Frequency response.

Flutter imparts a quivering quality to sound, and is especially noticeable on sustained notes. It is the result of rapid variations in the speed of a turntable or tape-transport mechanism. Although all record and tape players have some measurable flutter, it should not be audible on musical material.

Four-channel sound, also called quadraphonic sound, is a way of reproducing music with four separate sound channels to replicate sonic ambience from the rear of the listener in addition to frontal stereo presentation.

Frequency modulation, or, more commonly, FM, is a method of radio broadcasting in which information is transmitted by varying the *frequency* of the broadcast signal. It is better suited to high-fidelity transmission of music than ordinary radio, which operates on a principle known as *amplitude* modulation, or AM. Broadcast standards in the U.S. are such that FM broadcasts have a wider frequency range (up to 15,000 Hz, as compared to about 5,000 Hz in ordinary AM radio), freedom from static and noise, and a greater spread between loud and soft passages.

Frequency response describes how evenly a component responds to notes of different pitch. Ideally, a component should increase the loudness of all tones, from the highest to the lowest, by exactly the same amount. A frequency response varying no more than ± 3 db can be considered "flat" within the given frequency limits.

Gain, usually measured in decibels, means the amount of amplification and is sometimes applied to the volume control on an amplifier.

Ground, in audio terminology, usually refers to the metal chassis of the equipment. In the electrical sense, ground is the reference point from which most voltage measurements are made. When a technician measures the amount of voltage applied to a tube, for example, he attaches one wire from his voltmeter to the appropriate tube pin and the other to the chassis, or ground.

Heads are the parts of a tape recorder that apply the signal to the tape in recording and pick up the signal from the tape in playback. A separate head is sometimes employed to erase previously recorded material from the tape.

Hum is a steady low-pitched sound usually caused when a small amount of the 60-Hz voltage from the AC power line mixes with the audio signal. A certain amount of hum is inevitable, but if your equipment is in good condition, no hum should be audible with the amplifier controls set in normal playing position.

Impedance is an engineering term used to describe the degree to which a circuit impedes the flow of an alternating current. When two circuits or devices are electrically linked, their separate impedances must be matched for efficient transfer of energy. In hooking up an audio system, for instance, the impedance of the loudspeaker must be matched to the output impedance of the amplifier.

Input is the signal fed into a component—a radio signal, or the signal from a cartridge or tape head. Also, by extension, the connection or jack through which such a signal is fed.

Loudness compensation is a feature found in many amplifiers that compensates for the ear's loss of sensitivity to bass at low volumes. Music reproduced more softly than it would be heard in a concert hall sounds thin. Loudness compensation makes up for this by supplying more bass as the volume is turned down.

Mid-range is a rather arbitrary term for the central portion of the audible range. In current usage, the term refers to frequencies anywhere from about 500 to 7000 Hz. An amplifier (or speaker) that emphasizes the tones in the 2000- to 5000-Hz range makes the soloist seem closer to the listener, but the falsified tone color may lead to listening fatigue.

Mil, an engineering term meaning one-thousandth of an inch, is usually employed to specify stylus size. Common stylus diameters are .7 and .5 mil. The former can be used for all long-playing records (mono or stereo); the latter is best suited for newer stereo records.

Modulation, in a broad sense, is the process of altering an electrical signal so that it carries some sort of information—music, speech, dots and dashes, etc. In audio, the intensity of an otherwise steady electric current is varied in accordance with the sound-wave patterns of the program. In radio, modulation refers to superimposing an audio signal on the radio-frequency wave (the "carrier wave"). At the radio receiver, a circuit known as a detector then separates the audio signal from the carrier. Once the audio signal is recovered, it is amplified and fed to the speakers for reproduction of the original sound-wave pattern.

Mono (short for monaural or monophonic) refers to any record or sound-reproducing equipment designed for only one channel, as contrasted to stereo, which employs two separate channels.

Multiplex is a method of broadcasting in which two or more separate signals are broadcast simultaneously at the same frequency and from the same FM transmitter. This technique now forms the basis of stereo broadcasting, as it makes it possible for one station to send out both the left and the right channel of a stereo program.

Output describes the signal emerging from a component. Also, the connection through which it emerges.

Output power is the amount of energy the amplifier delivers to the loudspeakers; it is expressed in watts. One method of stipulating output power is in watts of *continuous power* (also called sine-wave power or rms power), which is the amount of energy the amplifier can generate continuously. Another measurement standard, called *music* or *dynamic* power, is less stringent. It allows for an amplifier's ability to exceed its continuous power rating for short bursts of musical sound, such as cymbal crashes. Still used occasionally is the older term *peak power,* which is usually the continuous power rating doubled. In rating stereo amplifiers, the output of both channels is usually added. For instance, a "30-watt" amplifier is usually an amplifier that delivers 15 watts per channel.

Phase, as the term is used in audio, usually refers to loudspeaker-hookup. In a stereo system, the two loudspeakers should be working in tandem, their cones pushing and pulling at the same time, rather than working against each other. When this condition holds true, the two speakers are said to be acoustically in phase. Conversely, when one speaker pushes forward while the other pulls back, the speakers are out of phase, which usually causes loss of bass and uneven sound spread. When speakers are out of phase, the situation can easily be corrected by reversing the connections to *one* of the speakers.

Pickup. See Cartridge.

Preamplifier is a section of the amplifier, sometimes called "preamp" for short, that contains the various controls to regulate volume, balance, treble, bass, and to select different program sources (records, radio, or tape). In most models, the preamplifier and the main amplifier (also called the power amplifier) are a single unit. However, in equipment designed for very high output power, the preamplifier and the power amplifier are separate pieces of equipment.

Quad—short for quadraphonic. (See Four-channel sound.)

Receiver. A component that combines a stereo tuner, preamplifier, and power amplifier on a single chassis, providing compactness and convenience along with a saving in cost. Only a record player and speakers need be added to complete the system.

Resonance is the name given to the tendency of any physical body to vibrate at one particular "preferred" frequency. The vibrations are greatest when the frequency of the applied force is the same as the natural resonance of the vibrating body. ("Natural resonance" is the frequency at which a body prefers to vibrate.) In audio, the electrical and mechanical resonances of the various components must be controlled so that they do not affect the tonal color of the music being reproduced. Equipment designers have therefore taken great care, for example, to place the natural resonance of phonograph cartridge styli above the audible range, and that of tone arms below the audible range. Resonance also has its analog in the electrical realm, making it possible to tune circuits to various frequencies, much as an organ pipe or a violin string can be tuned to a certain pitch. This is the principle by which radio stations are "tuned in."

Room acoustics play a large part in determining the overall sound of a hi-fi system. Shape, size, and furnishings of the listening room all influence the sound heard in the room.

Rumble, as its descriptive name suggests, is a low, rumbling noise produced by poorly built turntables or changers. It is caused by vibrations of the turntable mechanism that are picked up by the cartridge along with the signal on the record. In quality turntables, rumble is minimized by the use of properly balanced drive motors, shock mounts, and often by an elastic transmission between the motor shaft and the turntable rim (usually in the form of a plastic belt) that filters out motor vibration before it reaches the stylus.

Turntable rumble is measured, in decibels, against a standard-level tone played on a test record. The minimum requirement for high fidelity, as defined by the National Association of Broadcasters, is —35 db, meaning that the rumble must be 35 db softer than the test tone. (Note that unless turntable-rumble figures are specified as being derived according to the NAB standard, they cannot be compared with one another.)

Selectivity refers to the ability of an FM tuner to separate stations that are next to each other on the dial. In most large cities, the assigned frequencies of local FM stations are fairly well spread out over the dial, so that mutual interference is avoided. In some areas, however, FM stations received from different cities happen to fall too closely on adjacent frequencies. In such special locations, good selectivity becomes an important tuner specification. Selectivity is expressed as the number of decibels by which the signal from an interfering station in a nearby channel is reduced.

Sensitivity describes the ability of a tuner to pull in weak and distant stations. If you live in an urban area near all the FM stations you want to receive, sensitivity is not of great importance. But if you are located in a fringe reception area, a highly sensitive tuner can make the difference between satisfactory and poor reception. Sensitivity is always stated in relation to *quieting,* which refers to the ability of the tuner to strip off static from the incoming radio signal so that an interference-free audio signal emerges at the tuner output. If the specification reads: "3 microvolts sensitivity for 30 db of quieting," it means that the incoming signal picked up by the antenna must be at least 3 microvolts strong if the noise is to be quieted to a level 30 decibels below that of the music. The Institute of High Fidelity (IHF) suggests that all sensitivity ratings should be based on 30 db quieting. This standard is known as "usable sensitivity," or "IHF sensitivity," and most manufacturers observe this norm. Keep in mind that, in sensitivity specifications, the lower the figure, the higher the sensitivity.

Separation (rated in decibels) describes the ability of components to keep right- and left-channel signals apart. In poorly designed equipment, lack of separation permits signals from the right channel to leak over to the left, and *vice versa.*

Signal is the electrical waveform representing sound.

Signal-to-noise ratio (sometimes abbreviated S/N) expresses the relative amount of interference with the signal in a sound system or in any one of its components. In the language of electronics, "noise" is any kind of unwanted signal that intrudes into, or interferes with, the desired signal. In high fidelity, noise takes many forms: the rumble of a turntable, the hum of an amplifier, the hiss of a tape recorder, or atmospheric "static" superimposed on a radio signal. Perhaps the most consistently unappreciated pleasure of high fidelity is that all these forms of noise are held to a minimum by good equipment, and that the music emerges from a silent background. The signal-to-noise ratio is expressed as the loudness difference (in decibels) between the desired signal (usually measured at the equipment's full rated ouput under test or at some other standard value) and the interfering noise. In amplifiers, for instance, a specification reading "hum and noise −60 db" means that hum and other noises are at a level of 60 db below the desired signal reproduced at a given output level. A rating of −60 db is good—the higher the figure, the lower the noise. The signal-to-noise figures at high-gain inputs (such as tape-head or phono preamplifier) will always be worse than those of lower-gain inputs, such as tuner or auxiliary.

Solid state, in electronic parlance, is not a voting pattern but another way of saying "transistorized." It means that the equipment in question has no tubes (which contain a vacuum), and that its circuits use transistors and semiconductor diodes, which are solid throughout. A semiconductor, by the way, is a type of material that, electrically speaking, is neither fish nor fowl. There are a number of these materials, halfway between conductors and insulators, and they are the stuff of which transistors are made.

Stereo is a method of sound reproduction employing two separate channels to convey a three-dimensional sonic replica of the original performance. Stereo conveys not merely the sounds of the performance, but also a sense of the relative location of the various instruments in the orchestra and of the total acoustic environment in the studio or concert hall. Stereo is not something different from high fidelity, but is a further extension of the art of hi-fi sound reproduction.

Stylus is the modern term for what used to be called the needle in old-style phonographs. Unlike the short-lived steel needles used with old 78-rpm records, the modern diamond-tipped stylus is precision-ground and flexibly suspended so that it easily follows the contours of the record groove without damaging the disc or distorting the sound. (Also see Cartridge, Compliance, and Mil.)

Tape deck. A tape recorder designed to be permanently installed as part of a sound system, as contrasted to a portable tape recorder. Because it always plays through the amplifier and speakers of the main sound system, the tape deck usually has none of its own.

Tape transport. A tape-playing device without recording facilities, intended for playback only through an external amplifier and speakers.

Tone arm. The part of a record player that holds the cartridge and guides it across the record. It is precision made to move with minimal friction and to maintain correct stylus pressure on the record.

Tone controls are usually two, but sometimes four, separate controls that adjust the treble and bass response of the stereo channels. The controls enable the listener to boost or cut (weaken) high or low notes to attain pleasing musical balance.

Track. Up to four separate recordings lie side-by-side on the same tape—each of these is called a track. On standard four-track tape, each track occupies slightly less than one-fourth the width of the tape. When track is used as a verb (to track) it means the ability of the stylus to follow the record groove accurately.

Tracking error is an expression describing a less-than-optimum angle of the phonograph cartridge with respect to the record groove as the arm glides across the disc. Ideally, the cartridge should always remain perfectly tangent to the groove; practically, this is impossible because the arm does not move across the record in a straight line but in a slight arc as it swivels on its pivot. The deviation from the position of true tangency at any point on the record is called the tracking error. It is expressed as the angle between the true tangent to the record groove and the length-wise axis of the cartridge. The geometry of a well-designed tone arm—its curves and dimensions—is carefully calculated to reduce the tracking error to a minimum and to keep it minimal all the way across the record.

A low tracking error reduces distortion, particularly at the inner grooves of a record, where the mechanical problems of reproduction are particularly aggravated by the smaller arcs encountered.

Tuner refers to the section of a component system that tunes in radio broadcasts so that they may be played through the amplifier and speakers of the sound system. In contrast to ordinary radios, tuners are engineered for high-quality reception. They are available for all forms of broadcasting: AM and FM as well as stereo FM.

Tweeters are loudspeakers that specialize in reproducing high notes. Since the physical requirements for producing high notes differ from those for producing bass, audio designers favor a division of labor in speaker systems, entrusting the top notes to tweeters while the lower range is handled by woofers (see below). Sometimes there is a three-way division of frequencies, which requires the use of a special mid-range unit also. By thus dividing the sound spectrum between two or more speakers, treble and bass are kept from interfering with each other, and intermodulation distortion within the loudspeaker is avoided. To a great extent, the quality of the tweeter determines the over-all character of sound in a speaker system. A good tweeter must be free of spurious resonances (response peaks) that would cause a shrill or otherwise harsh sound. Moreover, it must scatter the highs broadly, spreading them evenly throughout the room. Wide-angle treble dispersion adds greatly to the stereo effect and to openness of the sound. Various methods are used in tweeters to achieve wide dispersion: dome-shaped diaphragms, flared horns, or sound-deflecting structures mounted in or in front of the tweeter cones.

Watts are the basic units by which electric power is measured. For instance, when you ask for a 100-watt light bulb, you are using the term to describe the power consumption of the bulb—the amount of electricity it uses. In audio, however, the term is most often used to specify not the amount of electricity needed to keep an amplifier running, but the amount of audio power the amplifier is capable of feeding to the loudspeakers. This wattage is known as an amplifier's power output.

Woofers are loudspeakers specifically designed to reproduce the low-frequency notes of the audio range. These bass speakers are bigger and heavier than tweeters because size and weight help them to operate more efficiently at low frequencies. A high-quality woofer should be able to reproduce the lowest notes of the orchestra (about 35 Hz) without difficulty. A loudspeaker system's ability to handle even the deepest notes without faltering adds a special feeling of depth and warmth to reproduced music. A woofer should be free of frequency doubling—the adding of gratuitous harmonics to the fundamental notes, thereby giving the bass a false coloration. A good woofer should also have good transient response—i.e., reproduce heavy bass notes sharply and clearly without booming or blurring. In order to do this, a woofer must be mounted in a properly matched enclosure, for without an effective enclosure even the best woofer will not sound good.

Wow is a slow waver in pitch caused by an unsteady turntable or tape speed. Wow can occasionally be cured by replacing the worn rubber drive parts in a defective turntable or tape deck. However, in poorly made components, the ailment is endemic, and the only cure is to replace the entire unit.

Index

About the Author

HANS FANTEL traces his interest in music and sound reproduction back to the age of four and a half when he could first reach the handle to crank up his father's phonograph at his home in Vienna. Since coming to the United States as a young refugee from the Nazis, he has followed the development of modern high-fidelity techniques from their inception during the period following World War II. He has contributed many articles on audio to such nationally respected publications as *The New York Times, Popular Science, Popular Mechanics,* and *Opera News.* As a contributing editor of *Stereo Review,* he has been one of the most widely read authorities in the field of sound reproduction. He currently serves as consultant on audio to the *Reader's Digest* Record Club.

Mr. Fantel has also written many magazine articles and several books on medical and scientific subjects as well as historical biography. His latest book, *The Waltz Kings,* is a history of the Viennese waltz and of its creators, the Johann Strausses, father and son. It is a selection of the *Reader's Digest* Condensed Book Club and the *Saturday Review* Book Club. At present Mr. Fantel divides his time between New York City and a hilltop house in the Berkshires in Massachusetts.